RIDDLED
WITH
LIFE

ALSO BY MARLENE ZUK

*Sexual Selections: What We Can and Can't
Learn About Sex from Animals*

MARLENE ZUK

RIDDLED
WITH
LIFE

FRIENDLY WORMS, LADYBUG SEX,
AND THE PARASITES THAT
MAKE US WHO WE ARE

HARCOURT, INC.

Orlando ❀ Austin ❀ New York ❀ San Diego ❀ Toronto ❀ London

www.HarcourtBooks.com

Library of Congress Cataloging-in-Publication Data
Zuk, M. (Marlene)
Riddled with life: friendly worms, ladybug sex, and the parasites that make us who we
are/Marlene Zuk
p.   cm.
Includes bibliographical references and index.
1. Pathology. 2. Human ecology. 3. Medical parasitology. 4. Human evolution.
5. Host-parasite relationships. 6. Adaptation (Biology) I. Title.
[DNLM: 1. Adaptation, Biological. 2. Disease—etiology. 3. Evolution.
4. Host-Parasite Relations. QH 546 Z94r 2007
RB112.Z85 2007
616.07—dc22   2006028642
ISBN-13: 978-0-15-101225-1

Text set in Requiem
Designed by April Ward

Printed in the United States of America
First edition

A C E G I K J H F D B

CONTENTS

RIDDLED
WITH
LIFE

INTRODUCTION

7:05 A.M. Get out of bed, brush teeth, gargle with Listerine. Wonder if the latter is really necessary, but killing 99 percent of germs can't hurt. Get dressed. Consider taking allergy pill, decide to wait and see.

7:45 Place bowl of cereal in front of two-year-old son. Watch son promptly fling spoon onto floor, where dog licks it enthusiastically. Son reclaims spoon, puts it back into bowl. Grab spoon and bowl, throw out cereal, get clean bowl and spoon, start again. Comfort screaming son and barking dog. Son stops screaming, starts coughing. Remember that son needs one more vaccination before summer. Dog stops barking, starts scratching. Remember that dog needs flea treatment.

8:45 Edge away from sniffling commuter on subway. Use end of sleeve to hold rail. Read newspaper article about potential threat of bird flu.

10:00 Use antibacterial wipes at work on telephone and desk. Listen to coworker's complaints about hay fever, and reconsider allergy pill.

NOON Go to new sushi place downtown for lunch. Make sure that the fish looks really, really fresh. Avoid dipping into communal bowl of hot sauce.

3:00 Read long e-mail from sister about perils of Internet dating. Latest complaint is that a guy who seemed great on the Web just doesn't smell right. It's not body odor, it's . . . well, she can't explain it. Sigh; wish sister would stop being so picky. Own husband was an athlete, is starting to develop paunch and has high cholesterol and blood sugar that's edging closer to diabetes. Think about what to have for dinner that he and son will eat.

5:15 Repeat subway vigilance, with fellow commuter now actually spitting into tissue, which then goes onto floor. Recoil. Wash hands with antibacterial soap at return home. Contemplate dinner.

6:40 Serve fish, though mercury is a concern. At least it won't have mad cow. Or bird flu. Husband is fine, but son won't eat fish, although it is white, his dietary color of choice. Worry that he will enter adulthood still refusing all foods besides pasta.

10:00 P.M. Watch news about AIDS in women of South Africa. Wonder how cousin's friend, who is HIV-positive and has been on a multi-drug cocktail for three years, is doing. Sneeze twice. Tomorrow, an allergy pill. Definitely.

It is no wonder that all of us feel under siege from the threat of disease. From headlines to television commercials, everything seems to be a war these days. In addition to the military conflicts around the globe, we are supposed to fight poverty, crime, and obesity, not to mention global warming and rising interest rates. Life is full of enemies, and to deal with them we need weaponry, ever more sophisticated and deadly: new laws, new diets, new economic plans. Nowhere is this warlike mentality more apparent than in our attitudes about disease. All day long we make decisions about strategy and battlefronts: the mouthwash, the hand gel, the vaccine, the flea drops.

No enemy is more feared, and the aggression toward it is unquestioned. We are fighting the invading microscopic foe, and we use drugs as if they were missiles directed at the combatants. It's ironic that our most powerful governmental physician is called the Surgeon *General.* Our lives are full of warnings about disease. We hear the message in ways large and small, from our earliest days, with the fearsome maternal admonition: Don't Touch That, You Don't Know Where It's Been. Sound bites and book titles blare warnings about flesh-eating bacteria, about a global flu pandemic, or unnamed but terrifying new plagues that threaten to devastate us all. The obvious solution is to fight back, and so we use a lot of warfare analogies in our discussions of disease. We battle the enemy of AIDS or Ebola or flu, we have killer T cells in our immune systems and worry about the bacteria invading (and we do say invading) our kitchens.

As long as you have your health, we say, everything else is secondary. As long as the baby's healthy, say expectant parents, there's nothing to fear. Anyone who has suffered from a major illness knows that the rest of life fades into the background during the course of the disease. The philosopher Lao-Tzu pointed out that health is the greatest possession, and Virgil supposedly said the greatest wealth is health (whether he was consciously plagiarizing from the earlier scholar is unclear). According to an Arabian proverb, he who has health, has hope; and he who has hope, has everything. The natural conclusion is that if we could eliminate disease, life would be paradise.

But stop for a moment. What if we didn't think about disease this way? What if we could relax our vigilance, at least some of the time? What if disease is not our enemy? What if it is simply a force we can learn to live with, like gravity? After all, gravity makes our skin—and other body parts—sag as we age, and our bones break when we fall. A weightless environment would require no wrinkle creams or Botox, no admonitions against standing too close to the cliff's edge. We could float through our world like fantasy characters. Yet we know that the bodies of astronauts too long in space begin to deteriorate, their muscles atrophying. We just aren't made to live in an environment without gravity. Its consequences may not all be wonderful, but we evolved with it. Disease is not as benignly neutral as gravity, but it has been part of life for longer than apples have been falling from trees.

What happens if we think of parasites not as enemies or friends, but as members of our family? We do not choose to have them, but our lives are unimaginable without them, and for better or worse, they have made us who we are. Some people have better luck, or less dysfunctional families, and some have worse, but no one grows up without their ancestors' influence.

Similarly, every kind of living thing gets sick. Always has, and always will. Sickness is our most persistent and ubiquitous threat, as it is for all other species, plant and animal. Both animals and plants get sick, with worms that enter their guts or speed through their stems, with bacteria that fester in wounds or choke their breathing passages, with fungus infections that wrinkle their leaves and rot their skin. One-celled organisms get sick. Even bacteria get sick, with viruses that specialize on them alone.

Disease is not merely ubiquitous. It is normal. It is natural. It is even essential. Illness has shaped all living things for millions of years, and life as we know it—we, as we know ourselves—would not exist without disease. According to an increasingly accepted theory, disease-causing agents contributed to the evolution of complex life itself, from its primordial start as a few chemicals casually sliding together like raindrops on a windowpane. Each of our cells contains tiny organs, subcomponents that produce en-

ergy or process fat molecules like microscopic livers or lungs. Biologists now believe that these organelles, as they are called, arose
from tiny primordial parasites that slipped inside a larger cell,
eventually taking over its machinery. They changed over the millennia from parasites to partners, enabling life to become more
elaborate. Eventually the cells banded together to make organs
and tissues, and organs and tissues united to form bodies. Those
bodies, in turn, continued to be invaded by other organisms,
small ones like bacteria or viruses as well as larger ones like
worms. And so we exist, organs and all, ever vigilant against the
dangers lurking in our kitchen sponges.

The males (in our species and others) do not think to credit
disease with their very manhood, and yet that too can be laid at
the feet of pathogens, which cleaved us from an early evolutionary asexual stage into two sexes, with all the turmoil that followed. Disease has gone on to influence how we choose our
mates, how we produce and protect our children, the reason we
distrust those we believe to be strangers. We may be attracted to
each other because we see, however unconsciously, a partner who
would be more resistant to disease.

This is not to say that disease is good, that parasites are our
friends, or that we should espouse some New Age "embrace your
illness" credo. Neither is it to suggest that illness is merely a state
of mind, an expression of weakness, and that if we understood
nature or were in harmony with the universe we would never
become sick. Illness exists, and it is a bad thing, but it cannot be
discounted as an aberration. It is not the exception to a normal
state—it *is* the normal state.

The British zoologist Gerald Durrell began his career as an
animal collector for zoos; his many popular books include *My
Family and Other Animals* and *Menagerie Manor.* Like many children
interested in animals, I loved his books, but a rather disturbing

incident described in *The Bafut Beagles* has stayed with me ever since I read it. Durrell had received a young Putty Nose Guenon (*Cercopithecus nictitans*), a kind of monkey with a rakish white spot on the face, from a hunter in Cameroon, and was getting it ready for transport back to England. The routine included a health check, and as was generally the case with wild-caught specimens, the animal was riddled with parasites—lice and mites in the fur, chiggers under the nails of the fingers and toes, festering sores from fly larvae burrowed into the skin, threadlike worms that writhed in the animal's droppings.

Durrell used the monkey's entirely unremarkable infested condition to point out that wild animals are not exactly living a healthy and wholesome existence in their natural habitats. He was particularly impatient with a rather twittery young woman who came to see his jungle camp and claimed to have a special interest in animal welfare. He acerbically observed, "I always think it rather a pity that people don't learn more about the drawbacks of life in the jungle before prating about the cruelty of captivity." I take a rather different moral from the story. If we find an animal with a tapeworm inside, with fleas burrowing into the fur behind its ears, or we look at a leaf with a tunnel mined inside by a chewing caterpillar, we conclude that the specimen is imperfect, that it doesn't look the way it is "supposed" to. If we wanted to photograph an example of any living thing, whether it is our dog or our potted geranium, we would want it portrayed unsullied, without bite marks or flesh dotted with the maggots of botflies or the eggs of worms. These perfect specimens, though, without blemish or infestation, are a fantasy, like the photographs we stage by waiting for the other tourists to leave, so we can make it seem to our friends back home that we were the only visitors to the Great Wall of China, or the Grand Canyon.

We are not wrong to be repulsed by disease; our ancestors

have instilled in us a deep horror of parasites that has served us well for thousands of generations. But we are wrong to think that an organism with these co-inhabitants is abnormal or unnatural. Life is naturally tattered, infested, bitten off, bitten into. The stem with a broken leaf, like an animal with lesions on its internal organs or less-than-glossy feathers, is more normal than its unscarred counterpart. An unblemished animal—or person—is idealized and fictional, like the advertisements showing a solitary traveler at the Eiffel Tower. It doesn't really exist except in our imaginations. Disease is part and parcel of how we are supposed to look, of how we are supposed to live.

I am an evolutionary biologist; I study the ways that animals have changed in response to their environments over the long arc of history. I see the marks of natural selection in almost everything about us—the color of our hair, our love of sweet fruit, the sweat that dots our brows. When the sun is hot, evaporation of our sweat cools us, allowing us to function at high temperatures. The sun presented a problem, and our bodies overcame it. End of story.

Diseases are different from the sun. The sun just sits there, glowing in the sky, with its eventual burnout and demise so far into the future that for all practical purposes—global climate change and gaps in the ozone layer notwithstanding—a defense against heat today will protect us tomorrow, and next year, and in the millennia to come. Not so for diseases, at least those caused by living things like bacteria or viruses. The sun will not get hotter if we sweat more or turn on the air conditioning. This is an obvious point, but it has enormous ramifications. Because organisms that cause disease are themselves alive, with their fate intertwined with our own, diseases can react back, circumventing our efforts to evade them. Unlike our story with the sun, the story with disease is never-ending.

Say that a kind of bacteria attacks the lungs. People with lungs that resist the invasion, maybe because the mucous inside is harder for the bacteria to penetrate, are more likely to produce children, perpetuating the genes for this superior mucous, than those who succumb. Bacteria that are adept at surviving in such mucous will outreproduce their less capable bacterial relations, and that in turn gives an advantage to lungs that resist the bacteria in a different way, perhaps by producing a chemical hostile to their growth. Point, counterpoint. No matter what we do, we are unlikely to get rid of bacteria completely, but conversely, they are unlikely to circumvent all of our defenses. We call this kind of interaction coevolution, meaning that the disease and its host evolve together, just like bees evolve to detect the flowers they pollinate. Of course, flowers need to be pollinated, while we could argue, wistfully, that we do not need to become sick. But the pathogens are with us nonetheless. And because every move we make is matched by one of theirs, we have a tight if reluctant partnership. It has been this way from the beginning of life itself. So every step we have taken, evolutionarily speaking—from the primordial ooze to land, up into the trees and back down again— has been accompanied by a clinging retinue of parasites.

What does looking at disease in this way mean? For starters, it means we can ease up a bit on that siege mentality so many of us take when confronted with illness. We do not have to see parasites as arising, unnatural and unbidden, in a different guise whenever we stamp them out. According to that popular and panicky view, West Nile virus replaces bubonic plague, to be replaced in its turn with AIDS or bird flu. We must remain ever vigilant to resist the latest onslaught, that philosophy says. Many of the doom-and-gloom books about emerging diseases feed this idea, with their warnings that we are in imminent danger of succumbing to the latest threat. Any discussion of parasites and disease is adversarial, resolute, warning us not to become complacent in the fray.

Though I agree that we take our unusually disease-free lives for granted in a world with modern medical advances, and that we often underestimate the power of parasites and infectious diseases, I take a different approach. I think we can lay down our arms and be better for it, maybe even healthier. The headlines about Ebola, avian flu, malaria, or antibiotic-resistant bacteria are alarming, as they should be. But we need more than alarm. We need appreciation.

People take courses in wine or music appreciation, of course, but it is more unusual to think about appreciating disease. Yet the first definition of appreciation in the *American Heritage Dictionary* is "Recognition of the quality, value, significance, or magnitude of people and things." Appreciation as an expression of gratitude comes later. We need to recognize the quality and value of illness in our lives. Think of this book, then, as a disease appreciation course, not in the sense of enjoying disease or making pets of our parasites, but of recognizing its imprint on every aspect of our lives.

Disease appreciation means backing off from those conflict metaphors. The analogy is correct, to a point; two parties with clearly different goals are in conflict. But this metaphor lulls us into the mistaken belief that it will be possible, at some point and perhaps only after great effort, for one party to win, once and for all, so that disease is vanquished forever. Nothing, however, could be further from the truth. We will never win, in the sense of living completely disease-free lives, no matter how much antiseptic we apply or how many swamps we drain. Our lives have been intertwined with those of parasites for so long that eliminating them, even if it were a realistic option, would not be a victory but a defeat that harms us even more than them. What we should hope for is not victory, but reconciliation.

Because infectious diseases are a delicate give-and-take between two parties, we must guard against a false complacency when victory has seemingly been achieved. Antibiotic resistance is a case in point; it is no accident that many types of bacteria can now shrug off the effects of previously powerful drugs like penicillin. Any biologist since Darwin will tell you that a few individuals in a population, whether it is a population of bacteria, clams, or puppies, will be able to fend off any particular threat, simply because individuals vary and by happenstance one or another will have, say, a darker shell less visible to predators, or a cell structure that is impervious to antibiotics. Because the bacteria reproduce with blinding speed, the slight advantage gained by the resistant individuals is perpetuated among their offspring, and soon they dominate the population. Bacteria become resistant exactly the same way that fish descendents became able to walk on land, and providing an environment pervaded by weakly antibacterial substances in dishwashing soaps, hand creams, and cutting boards just hastens the process. And so we have created a world in which syphilis, tuberculosis, and even a simple infection with staphylococcus bacteria now disregard the drugs we thought would vanquish them forever.

The idea of disease-as-family can also help us rethink the way we devise cures for our symptoms. For example, when we get the flu, our bodies produce a variety of symptoms: We cough, run a fever, or feel weak or tired. Normally we try to get rid of all these symptoms, so that we feel better, at least in the short run. But if instead we view that illness as an interaction between two active participants, the host (us) and the disease organism (the virus, worm, or bacteria), then we can see some of those symptoms as defenses that we are mounting against the disease. Fever, at least at relatively low levels like those accompanying most colds or flu, is a good example of this, and one that I will discuss in more detail later; the slight increase in body temperature seems to stifle the development of some harmful microorganisms, and cold-blooded animals like lizards will even seek out hotter temperatures if they are sick. Other symptoms may represent a pathogen's attempt to be spread, such as sneezing, which spreads virus-laden respiratory droplets. We do best in suppressing them, but the point is that not all symptoms are alike.

We are only just beginning to understand how our bodies have evolved with parasites, and how the overly cleansed world into which we now proudly place our newborn babies may be a deceptive haven from illness. Don't Touch That, You Don't Know Where It's Been. And yet—allergies and childhood asthma are at an all-time high, particularly in industrialized parts of the world where antibacterial soaps abound and children never play barefoot in the muck by the river. Autoimmune diseases like inflammatory bowel disorders are on the rise as well. Scientists are beginning to speculate that the absence of challenge to the immune system, which defends us against foreign intruders like bacteria and worm eggs, may have unanticipated repercussions. Like bored children, immune system cells may start working mischief, attacking the cells of the very body they inhabit or initiating an elaborate assault on an innocuous pollen grain or mote of dust. Having evolved with parasites, could we suffer when they are gone?

When people talk about parasites and their influence on us, they usually mean horror-movie events, with hapless victims being taken over by ruthless pathogens and becoming mass murderers in their manipulated efforts to spread the parasite to new hosts. I have always had a sneaking sympathy for the parasites in these situations. What happened to free will in these films? And why doesn't the parasite ever get any credit for its success? Just as in nature documentaries that portray the gazelle or rabbit as the sympathetic victim, but rarely show the predatory lion or wolf as hero, we somehow always identify with the attacked rather than the attacker, an irony given our own place in the food chain. One critic reviewing the 1975 movie *Shivers,* about a rogue parasite that gives its victims an insatiable sexual appetite, did manage to see the parasite's point of view, suggesting that the film's "themes include . . . a fascination with the fragility and mutability of the human body, and the radical possibilities of transcending evolution by using science to drastically alter our bodies and minds."

In reality, we hardly need to use science to drastically alter our bodies and minds—the parasites do just fine at that, thank you, and in ways mere filmmakers could never have imagined. It isn't a technological fantasy that allows the human body to mutate, its wants and needs twisted by the parasite inside—it's real life.

In the end, I do not believe that disease is on the brink of overcoming us all, or that we have to remain in a state of panic at all times. Sweeping epidemics are not the inevitable price we pay for having engineered our environment and instituted technology, nor must we redouble our efforts to slaughter every microbe we discover. Instead, we can and should reconsider our relationship with disease, in a way that gives great hope for the future. We will never discover a cure for all that ails us, and new diseases will always arise and spread. But if our side can't win, neither can theirs. Disease is everywhere. We can't give it up any more than we could give up any other member of our family.

CHAPTER 1

WHY DOCTORS NEED DARWIN

What do we do when we get sick? The details, of course, depend on the illness, but the immediate and obvious answer is that we try to make ourselves feel better. A cold or bout with the flu usually spurs us to the pharmacy for cough lozenges, aspirin, and decongestants, while an upset stomach means finding a remedy for running to the bathroom. Alleviating the symptoms may not be a permanent cure, but it can't do any harm, right?

Not necessarily. Although I dislike the analogies between war and disease, it does help to think of symptoms as the sounds, bullets, and urgent messages coming in from the front, with no indication of whether they are from friend or foe. Wiping them out indiscriminately risks destroying our own troops with friendly fire. In some ways, the Calvinists were right—suffering can be good for you, though the more modern application of this idea suggests that it is not the soul but the body itself that benefits from pain. The trick is to figure out when that is the case, and when there is no virtue in misery.

And what about diseases that do not result from infectious agents like bacteria or viruses, but seem to be a flaw in the factory model? Although some aspects of the human body can seem like a miracle of precision in design, our knees and backs give out and childbirth is not exactly a walk in the park. We have wisdom teeth that cause problems because there is no room for them in our jaws. Genetic diseases like cystic fibrosis persist in the population.

We scientists always point to the amazing adaptations of organisms, whether it's the design of our eye or a lowly moth's wings that mimic leaves so perfectly that they have built-in tatters and stains from bird droppings. If nature does such an incredible job with camouflaging a mere insect, why has she dropped the ball with our vertebrae?

A relatively new field, called evolutionary medicine or Darwinian medicine, takes a radically different perspective on health and illness to answer these questions. First articulated by psychiatrist Randy Nesse and evolutionary biologist George Williams, Darwinian medicine places diseases and defects in an evolutionary framework to make sense of the apparent mismatch between the way our bodies often work and the way we would like them to: Natural selection may not have produced diseases like diabetes or arthritis directly, but it has made bodies vulnerable to them for a variety of reasons.

DESIGNED BY A COMMITTEE

Back and knee problems are the source of enormous amounts of pain, and they cost society millions of dollars in treatments ranging from anti-inflammatory drugs to surgery, lost work time, and insurance payments. Lower back pain alone is said to afflict nearly half of many adult populations at one time or another within a given year. It is not confined to industrialized or even modern-day populations. Evidence of arthritis of the vertebrae and joints can be found in prehistoric hunter-gatherer skeletons, and back problems are even common in less technologically advanced populations, although the prolonged sitting of a cubicle-centered lifestyle seems to exacerbate the malady. In search of the root cause, many scientists point to the toll that it takes for an originally four-footed mammal to get up and spend all of its time on its hind legs. Other apes are bipedal, or moving on two legs, for short periods, but our way of walking and running exclusively on two feet is ours alone. The evolution of bipedalism, thought to have arisen over 4 million years ago, may have freed humanity's hands for stone tools, food baskets, and mobile phones, but it also caused a shock to the skeletal system from which we have still not recovered. Anthropologists speculate that our joints would have to be too large for efficient locomotion if they were also stable enough to bear our body weight, resulting in the current compromise of aching backs and knee braces.

Bipedalism also seems to be responsible for the change in pelvic anatomy between humans and our apelike ancestors. The short, broad pelvis of humans supports the torso and the gut, which is a good thing when you are walking upright, but it also made the birth canal narrow, which is not such a good thing when you are trying to have a baby. In addition, brain and head size seem to have increased quite rapidly in humans over the last million years or so, and while much of the growth of the infant brain takes place after birth, the size of the head of a newborn baby is still perilously close to the size of the passage it emerges from, something that causes more complications at delivery in people than in other primates. Childbirth is still a challenge for other mammals, but it is virtually always more straightforward in all animals than it is in humans. Interestingly, humans also seem to differ from most mammals in their apparently universal need for emotional support from others during labor and delivery; birth is unaccompanied in most mammals, but women from many if not all cultures seek out other people during childbirth. Anthropologist Wenda Trevathan suggests that the shift in the presentation of the baby during delivery to exit the birth canal face first, another result of our bipedal skeletons, also led to an increased risk of mortality if births were unattended. Hence, she argues, early humans were safer if birth became a social, or at least a supported, affair, something she suggests modern birthing practices would profit from understanding.

Regardless of these travails associated with walking upright, walk upright we do, and Darwinian medicine suggests that problems like complicated births and backaches may simply be part of an evolutionary legacy. Nesse says, "We think the body is designed for health, but it isn't. Natural selection maximizes reproduction, not survival or health." This is a key point. Of course a certain level of good health is requisite for functioning, so that the truly enfeebled are at a disadvantage, but as long as an organism can reproduce and pass on its genes, all of its flaws will be passed on as well. François Jacob said that nature is a tinkerer and not an engineer; in other words, bodies got that way not through goal-oriented advance planning but by a series of lurching steps. Certainly a hypothetical mutant with a good back and wider pelvic channel who also had more offspring could prosper, but if possessing such traits imposed too many constraints on locomotion, said creature might never arise, or if it did, wouldn't make it out of the starting gate.

Another problem in producing a perfect body is that evolution can only change things that happen at certain times of life. Say a bird can produce more eggs early on in life, because her body can turn calcium into eggshells more quickly than other birds can. That bird would be blessed with chicks over and above her relatively eggless relations. But what if the same gene that gave her all those eggs while she was young also caused her bones to become more brittle as she aged, resulting in the avian equivalent of osteoporosis, so that she couldn't fly? Many genes do have such multiple effects, and biologists are extremely interested in the consequences of their actions. Surprisingly, it turns out that as long as a gene's deleterious effects are postponed past the first peak of reproduction, and as long as that gene gives its bearer a sufficiently large advantage, such genes will be perpetuated in the population. This good-early–bad-late interaction is called negative pleiotropy, and it could be the evolutionary answer to why we grow increasingly decrepit in old age. It also means that imperfection in one body part is a regrettable but inevitable consequence of natural selection acting to perfect another. It's not that the good die young, it's that being good while you're young matters most.

GRASSHOPPERS, ASPIRIN, AND THE BEST DEFENSE

Pain and suffering are hallmarks of disease, and while philosophers and poets can wax eloquent about their benefits to the spirit, both physicians and their patients are united in wanting to eliminate them from the body. All forms of suffering are not created equal, however, and Darwinian medicine helps us understand the consequences of assuming that they are. Pain itself is a useful signal, of course. The few individuals born without the ability to feel pain lead very complicated and controlled lives and generally die at an early age. It is easy to understand why; we rely on pain to tell us when to move our hand from a hot stove or how far to bend a joint. But what about disease symptoms that range from annoying, like the itch of a mosquito bite, to debilitating, like the cough of pneumonia? What about the general malaise, the mopiness and lethargy that accompany a wide range of illnesses? Could they, too, serve a useful purpose?

In particular, what about fever, the ubiquitous partner of illnesses ranging from colds to malaria? Fever is the most common reason for parents to bring their children to the hospital emergency room, and millions upon millions of dollars are spent each year on fever-reducing drugs like acetaminophen, ibuprofen, and aspirin. Most parents believe that high fevers, those above 104° F (40° C), are dangerous, and can cause brain damage if left untreated.

But Hippocrates was a strong advocate for the beneficial effects of fever, believing that it burned off excesses of the humors or essences of the body, and many cultures around the world used to induce fever to treat disease; in at least one Native American tribe, a sufferer was placed inside the carcass of a freshly slaughtered horse to absorb the heat lingering in the body cavity. In 1927, the Nobel Prize for medicine went to Julius Wagner-Jauregg, an Austrian physician who had tried many cures for the fatal paralysis caused by late-stage syphilis. His breakthrough came when he deliberately infected syphilitic patients with malaria to induce high fevers; most of them showed striking disease remission, whereupon he cured the malaria with quinine. Wagner-Jauregg was not completely certain why his treatment worked, pointing out in his Nobel acceptance speech that the high temperature alone was not the sole mechanism behind recovery. He speculated that the fever activated some other component of the body's natural disease resistance, but had little information to support this suggestion, since the workings of the immune system were only beginning to be understood. More recently, malaria therapy has been suggested for the treatment of Lyme disease, some forms of cancer, and even AIDS, but it is viewed with considerable skepticism by the medical establishment.

High fevers got a bad reputation during the nineteenth century, when scientists who experimentally elevated the body temperature of laboratory animals by five or six degrees above normal showed that the animals then suffered brain damage and often died. Fever-controlling medications also began to be popular in the late 1800s, perhaps because most of them, such as aspirin, are also analgesics that relieve pain. Pain relief therefore became synonymous with fever reduction, but no one ever tried to determine if suppressing a fever alone would have beneficial effects. Then, in the 1970s, a physiologist at the University of Michigan medical school named Matt Kluger and two of his graduate students decided to examine fever, not in humans but in animals like lizards. Such animals are sometimes called cold-blooded, but in addition to being (in my opinion) needlessly pejorative, the term is inaccurate, since the real distinction between the birds and mammals on the one hand and fish, reptiles, amphibians, and invertebrates on the other is not the temperature of the blood but the way body heat is regulated. Heat can be manufactured and regulated either internally, as with the birds and mammals, or externally, as with the lizards, more appropriately called ectotherms, meaning "outside heat."

Such animals can maintain an impressively high body temperature if allowed to seek out warm places to bask, and one can ask whether the temperature they seek changes depending on their condition; in other words, do lizards give themselves a fever when they are sick, and if they do, does it help? Ectotherms are better subjects for asking this question than endotherms like humans because you can manipulate their body temperature by changing the temperature of the environment, rather than by administering drugs like aspirin that affect more than just fever.

Kluger and his students took desert iguanas and injected some with killed bacteria to mimic an infection. They then allowed the lizards to choose their body temperature by giving them a chamber with varying temperatures, so that the lizards could select the cooler end or the warmer one, like a cat snoozing in the sunny spot of a room. The bacteria-injected lizards chose a warmer place to rest than the uninjected ones, which caused them to have a higher body temperature than usual.

Such "behavioral fever" has now been demonstrated in a

wide variety of ectotherms, including alligators, frogs, fish, lobsters, scorpions, grasshoppers, and beetles. Experiments have shown that seeking high temperatures does indeed help cure the animals of disease, and that animals prevented from going to a hotter place are more likely to show severe symptoms or even die. The grasshoppers that show behavioral fever have been of particular interest because they are serious pests in many parts of the world, periodically sweeping fields in devastating clouds of millions. A promising defense has been spraying them with species-specific diseases such as bacteria or fungi; these are generally judged to be safer than standard insecticides because they do not kill helpful or innocuous insects along with their intended target. But the grasshoppers have their own means of defense against disease, and one experiment using such a fungal control agent showed that as long as the insects could move to a warmer area and generate a fever, they were still able to reproduce. Although this didn't mean that the fungus had to be abandoned as a means of keeping the pest numbers down, it did suggest that farmers in warmer climates, for example, might find the treatment less effective.

In mammals, fever also seems to aid in recovery, although as mentioned above the experiments can be harder to interpret. Rabbits, for example, can lower the numbers of disease-causing bacteria in their systems more easily if they develop a fever. Goats that were infected with a one-celled parasite and then given a fever-reducing drug died, while those left untreated had a mild infection and recovered. Other animals show similar responses. Artificially lowering the body temperature seems to interfere with the ability to fight off a variety of bacterial and viral infections in animals from crickets to fish and mice. Fever seems to be an ancient evolutionary response, stubbornly retained as animals changed from having fins to legs or wings, gills to lungs, scales to fur and feathers. This pedigree, along with the experimental evidence about the benefits of fever, led Kluger, along with many other scientists, to classify fever as a helpful, adaptive response of the body to illness, not the menace it became known as in the last century.

It is important to distinguish between a real fever, generated by the body in response to disease, and what physiologists call hyperthermia, or high temperature caused by external factors like heatstroke. For mammals, unlike lizards, simply raising body temperature is not enough, which is why sitting in a sauna when you are sick may make you feel good but won't really mimic the benefits of fever. Those nineteenth-century experiments on laboratory mammals were demonstrating heatstroke, not fever. (This also probably means the technique of stuffing sick people into the steaming entrails of a horse is not likely to be effective, which is probably just as well.) True fever involves a recalibration of the body's internal thermostat, so that instead of keeping things ticking along at, say, 98.6° F, the new set point is 100° or 101°. The fever response is complex and intricately orchestrated, as Wagner-Jauregg suspected, and modern scientists believe it regulates the production of cytokines, chemicals produced during the immune response that are responsible for the movements of white blood cells that carry out a variety of functions.

After all this feverish enthusiasm, one might wonder why the body's set point isn't simply a little higher, so that all the benefits of fever could be enjoyed on a daily basis. The answer is that everything has a price, and fever is metabolically quite costly. Keeping the body at a given temperature requires energy that must be extracted from food, and that energy could also be used for the multitude of other activities an animal must perform. Toads given a fever-inducing agent had higher metabolic rates than control toads when both groups were kept at a temperature equivalent to that sought by the injected toads, suggesting that mounting the fever response itself is expensive. In addition, some scientists have speculated that lower temperatures might help stave off aging by reducing the amount of toxic materials produced by the cells during an immune response, making it best to maintain as low a body temperature as we can get away with.

WORSE THAN THE DISEASE?

So what about that widespread use of drugs to lower fever in children? Practices are changing, albeit slowly. Some medical practitioners now warn against "fever phobia," the needless panic felt by many parents and health care providers when a child's temperature rises. A paper published in the *Bulletin of the World Health Organization* surveyed numerous studies on the use of fever-reducing drugs in children and came to the rather startling conclusion that they made no difference in the outcome of the disease, the duration of the symptoms, or even the comfort level of the children themselves. In one of the studies, parents were not told whether they were giving their children a potent drug or an inactive placebo (they agreed to this in advance). When asked which they thought the child had received after the sickness was over, the parents guessed right about half the time, exactly what you would expect by chance. A slightly higher degree of activity and alertness was noted in the children receiving the medication, but this was minor. The authors acknowledge that this is not the final word on the subject, but it does give food for thought.

Medical researchers have also debunked two commonly held misconceptions about high fever in children: that it can result in dangerous seizures, and that fevers from infection must be controlled before they reach a certain point, often 41°C (106°F), to prevent seizures and brain damage. Febrile seizures, as they are called, are certainly frightening to watch, but they tend to occur early in the fever process, rather than after fevers have mounted, and a small percentage of children simply seem to be prone to them; administering fever-reducing medicine does not stave off their recurrence. They also do not have permanent ill effects, and although parents are advised to notify the doctor if their child has one, they are not necessarily a cause for alarm. And while it is true that fevers over 106°F are potentially damaging, such high temperatures are virtually always the result of heatstroke or brain injury, not infection, and so fears of a cold- or flu-caused fever rising to this level are groundless. Michael S. Kramer and Harry Campbell, two child-health experts writing in a document for the World Health Organization, say, "One is left to conclude that the principal rationale for antipyretic [fever-reducing] therapy is to soothe worried parents and health care workers and to give them the sense that they are controlling the child's illness, rather than it controlling them."

This is ironic, since it is not so simple as us versus them. Fever, as a mechanism that activates the immune system to cure us of the pathogen, is a defense, a tool on our side, and the best way to control an illness is to leave the fever alone, at least some and perhaps most of the time. Certainly we can choose to suppress fever if we have a task to do that requires a more alert mind, but we need to be mindful of the price that exacts.

And it's not just fever. Other symptoms that bear reevaluating include coughs, which can rid the lungs of harmful matter; diarrhea, which likewise sends the offending agent rapidly through the digestive system; and even the behavioral accompaniments of illness such as sleepiness and lethargy. Many animals exhibit sickness behaviors like those of humans; they go off by themselves, refuse to eat, and do the zoological equivalent of watching daytime television. Some veterinarians and researchers suggest that these behaviors are more than the side effects of infection; they could be defenses that help speed recovery by allowing energy to be directed at healing. "Soldiering on" during a bout of flu may thus do more harm than good.

Reduced appetite during sickness is a particularly interesting symptom, because it may be linked to a mechanism for eliminating virus-infected cells from the body. Such illness-associated anorexia, as distinct from the eating disorder anorexia nervosa, is a hallmark of many diseases, including AIDS. Many doctors attempt to treat it and get the patient eating again, but Edmund LeGrand, a researcher with Johnson & Johnson, wondered whether the controlled starvation during illness might serve a function. In particular, he suggested that it might favor a kind of directed cell suicide called apoptosis. Apoptosis is a complex and orchestrated process, distinct from simple cell disintegration in the way that fever differs from heatstroke, and it targets cells that are already infected with virus particles, which can help control the disease. Food restriction makes this process happen more easily, and LeGrand speculates that the wasting seen during HIV infection might actually be beneficial. This is not to say that extreme weight loss and lack of nutrients during disease is helpful and should never be treated, but it is possible that the mild loss of appetite associated with some illness is therapeutic.

While leaving some diseases to run their course may be best, we are obviously not going to stop all forms of treatment, nor should we. Some manifestations of disease are defenses, but others are defects. Again, however, thinking about medications in light of our relationship with disease is useful. Our bodies' regulatory systems, which maintain everything from our internal salt and water balance to our flow of oxygen, tend to respond to any perturbations by attempting to get those mechanisms back to the way they were before. Drugs that mediate one of these regulatory systems, maybe by altering the production of a hormone, spark the regulatory system to fight back, so to speak, which makes the drug less effective.

Take, for example, the use of leptin to treat obesity. Leptin is a substance produced by fat cells; it helps regulate a variety of systems including appetite. The more body fat a person has, the more leptin he or she produces, and at least in laboratory rodents, that reduces appetite and promotes weight loss. But administering leptin to obese people doesn't help much, and in fact it turns out that such people actually have higher levels of circulating leptin than those who are not overweight. What's going on? The answer may lie in the evolution of the regulatory system. It is reasonable to suppose that humans evolved stronger mechanisms for gaining weight and keeping it on than for taking it off, since the risk of starvation was far more real than the risk of eating too many Oreos. Scientists suspect that under normal, or at least pre-Oreo, conditions, hunger levels went down when leptin levels increased. If we were starving, however, appetite was *stimulated* by low leptin levels and the brain was told to increase fat storage and maintain vital body processes, maybe even at the expense of reproduction. The problem is that weight loss in an overweight person can look like starvation to his brain, so his regulatory system compensates for the leptin drop by increasing appetite and encouraging weight gain. The system has no way of knowing that the new weight is healthier than the old one—recognizing obesity simply wasn't necessary in most of our evolutionary past. Any treatment of obesity needs to take this evolutionary heritage into account. Researchers are toying with altering the baseline levels of leptin, so that the system is short-circuited at an early stage, before the delusion that starvation is occurring sets in.

Randy Nesse advocates using the smoke-detector principle in deciding whether to treat a defense symptom. Smoke detectors are designed to be sensitive to any amount of smoke, even if that means having to listen to the earsplitting screech after burning toast, because the cost of such a false positive is much lower than the cost of the false negative of ignoring a smoldering cigarette in the bed. Similarly, the body's defenses err on the side of safety. Some defenses are all-or-none, like vomiting; we can't stop halfway, as it were. Others are graded, like fever, where the rise in temperature can be slight or marked. The all-or-none defenses should occur when the cost of responding is relatively low compared to the cost of doing nothing; expelling a poisonous food costs only the calories consumed and the energy in eliminating them, while leaving the toxin in the body might be very costly indeed. More graded responses should be more conservative, because it is easier to fine-tune exactly how high a temperature increases. This means that we have some leeway, particularly for the high end of the graded responses. In other words, going ahead and treating fever may not hurt, since the defense mechanism is oversensitive and fever may kick in at a level of illness where it is not essential to raise the body temperature. In general, though, if a symptom is caused by the host, rather than by the pathogen, we are best off leaving it alone.

DISEASE AND THE FLINTSTONE-JETSONS

The idea that somehow, somewhere, we have gone terribly wrong as we came down from the trees, or out of the African savannah, or when we stopped hunting and gathering for a living, or moved into cities, or started wearing stiletto shoes and eating Mars bars and hamburgers, has recurred in countless novels, philosophies, and self-help books. We idealize Rousseau's Noble Savage, we go on nature retreats, and we natter on about slow food and the joys of simpler lives. Nowhere is this more apparent than in the mountain of supersized dietary advice suggesting that we would live longer and weigh less if we ate more like our ancestors. Just which ancestors is not entirely clear, though one popular version, the Stone Age diet, suggests that we abandon dairy products and the grains and other cultivated starches that accompanied the development of agriculture and live off of lean meat and raw fruits and vegetables, kind of like Atkins but without a lot of cooking. As *USA Weekend* magazine put it, "Your body craves nutrients cavemen ate."

Superficially at least, this enthusiasm for the twigs and berries of yesteryear, or maybe yester-epoch, sounds like it should be supported by Darwinian medicine proponents, who maintain that there is a mismatch between the circumstances under which humans evolved and our present-day environment. Our Stone Age genes, as S. Boyd Eaton and Stanley Eaton say, "now contend with the realities of Space Age life." As a result, modern humans suffer from the diseases of affluence and civilization, such as diabetes, clogged arteries, and obesity. But a closer look reveals that not all Darwinians are created equal; the devil, even for Darwinians, is in the details. Just because our species evolved in a different environment does not mean—and Darwinian medicine does not necessarily say it means—that following the ways of the past is automatically going to free us from the illnesses of modern life.

The first question is just when this proverbial Stone Age happened, and what the transition from a nomadic, hunter-gatherer existence to a more settled agricultural life really meant in terms of disease. The Paleolithic, to use a somewhat more scientific term for the Age of Flintstone, is the period from about 2.5 million years ago to around 10,000 years ago, give or take a few thousand years. This is a hefty span of time, and humans weren't doing the same things throughout it. Scavenging, or eating the carcasses of dead animals left by (or stolen from) predators such as lions, was probably replaced by active hunting and the accumulation of wild plant material about 55,000 years ago. Agriculture and the domestication of plants and animals began a mere 10,000 years ago, apparently more or less simultaneously in several places around the world.

So how old does this make our genes? Although evolution can occur quite rapidly, over the space of only a handful of generations in some cases, by and large our genes have not changed very much since the origin of agriculture. This does not, however, mean that we have Stone Age (or Paleolithic) genes; most of our genes are far more ancient than that, and indeed we share a huge number of them with organisms as unlike us as fruit flies and sea anemones. The much touted 98 percent genetic similarity between us and chimpanzees is not all that meaningful in at least some contexts. We certainly spent longer as hunter-gatherers than as farmers, much less as computer analysts, but the significance of those relative time spans is not all that clear. This suggests that it's rather arbitrary to pay more attention to genes from the Pleistocene than, say, the Devonian period of about 350 million years ago, also known as the Age of Fishes, when some of our vertebrate ancestors were sucking body fluids of other fish for nourishment.

That said, the invention of agriculture was indeed a critical point for the evolution of disease in humans, and not just because it changed the human diet. Agriculture means a sedentary life with groups of people living permanently in villages or towns. Crops also support more people than hunting and gathering, and larger populations provide a reservoir for many infectious diseases. Measles, for example, requires a rather generous population base, with frequent immigration of new hosts, to survive, and it is thought that measles in humans did not exist until people began to form larger groups than nomadic hunter-gatherers could support. Settling down means using the same water sources and waste repositories over and over again, and many diseases are associated with inadequate sewage treatment. Settlement also allows disease-causing organisms to complete their life cycles; the parasitic liver fluke that causes schistosomiasis spends part of its life in snails in rivers and streams, and leaves the mollusk when infected people go into the water to bathe or drink, only to be released back into the water, and into another snail. Finding another person to inhabit would not be possible if the water source were not being reused. In addition, most farmers use manure as a fertilizer. This used to include human excrement (euphemistically referred to as "night soil"), providing another source of contamination with pathogens. Finally, domesticated animals, especially mammals like cows and pigs, help disease organisms pass between species. This is not an everyday occurrence, but the sheer amount of contact between people and the animals they herd or raise has sometimes led to diseases "jumping" from one species to another, like the swine flu that originated in pigs but mutated to find a host in humans. Avian flu is our most recent candidate. Although scientists are not uniformly convinced, HIV is thought to be another of these zoonoses (diseases that moved from animals to humans), as a similar virus is found in many of the monkeys used as food in Africa where the disease seems to have originated. The rats that carry the fleas infected with bubonic plague are also associates of settlements rather than nomads.

Becoming farmers thus gave us a whole host of new diseases, but that doesn't mean that if we just ate like the hunter-gatherers that preceded farmers we would avoid all our modern ailments. Even the idea that their diet is more natural is questionable. Certainly they eat a more varied diet; contemporary hunter-gatherers commonly consume more than one hundred different kinds of plants, while agricultural people tend to eat maybe a dozen. This breadth likely yields more nutrients than a more monotonous selection. Both hunter-gatherers and traditional agriculturalists, including virtually all humans up to a few centuries ago, consumed far more unprocessed food and thus fiber than most Westerners do today, and hence suffered far less from digestive disorders such as diverticulosis. The amount and kind of meat eaten by hunter-gatherers, a key point of contention among the Stone Age diet proponents, was variable, but certainly hunted animals tend to have less fat than domesticated ones.

What does this information tell us about what whether an old-fashioned diet could help us avoid getting certain diseases? First, a great many historical and evolutionary changes contributed to the kinds of diseases to which we now succumb. Becoming agricultural was only one of them, and the actual diet of farming people is only one among numerous risk factors associated with that 10,000-year-old shift in society. Second, it is not actually clear what period, or which ancestors, people are talking about when they bemoan how unhealthy a modern diet is and how much better off we'd be eating like Stone Agers. One magazine article advised against starting to lose weight in the fall or winter, because our ancestors had a predisposition to put on fat and store it for the long cold period. Better, they counseled, to start a diet in the spring. Hold on—which early human ancestors are these, the ones from Minnesota? I am not contesting the finding that people could find it easier to lose weight in temperate climates when it is warm outside, but that seasonality may have more to do with the lure of the couch in winter and the threat of impending bikini season in the spring, or perhaps the effect of daylight on circulating hormones, than with a mythical hibernating forbearer. Remember, humans evolved in Africa, spreading only gradually, and much more recently, into harsh temperate climates.

But the third conclusion is that just because there is a mismatch between The Way It Was Then and How We Live Now doesn't mean that it would be better to follow our ancient dietary instincts or trust that hunter-gatherers have some ancient genetic wisdom. The ancient-diet-is-best crowd also suggests that if we just listened to our bodies, they would tell us to eat the right foods in the right proportions, and those foods wouldn't be French fries and Snickers bars.

In fact, the opposite may be true. The reason we got into this nutritional and dietary mess in the first place may well be that we inherently like foods that were highly nutritious but in short supply during our evolutionary history. Seeking out sugar in the form of ripe fruit or honey gave our forbearers a good source of energy and helped avoid the plant toxins sometimes present in unripe fruit. As for craving the nutrients cavemen ate, the answer is yes—if you are willing to accept the fact that they would have been downing candy bars and soft drinks with the best of us. So the foods we want the most have never been the foods that are best for us, not because we are somehow out of harmony with our bodies, but because those candy bars weren't around for most of our past.

Going back to a pre-agricultural diet, rather than, say, an early agricultural one, or even a modern one bereft of quite so many Cheez Doodles seems rather arbitrary to me; the nutritionists are divided on the benefits of high versus low carbohydrate intake, and few advocate the complete elimination of entire large groups of foods. Besides, exactly which hunter-gatherers do we want to emulate? Aboriginal Australians ate far differently than Inuits, who in turn differ sharply from Kalahari Bushmen. One version of the back-to-the-Paleolithic diet advocates eating a lot of fish, claiming that this was part of our heritage, something the desert natives would have a hard time fulfilling. Saying, as Nesse and Williams do, that modern life is at odds with our evolutionary heritage is not to value one over the other. Darwinian medicine is not synonymous with promoting a return to the past.

It is also worth remembering that people vary in their sensitivity to increased amounts of salt, fat, and sugar, just as they do in many if not most other traits. Some people are simply unharmed by eating a lot of junk food, irksome though that may be to the rest of us. Until very recently, natural selection didn't weed out those who developed high blood pressure from eating a big bag of salty potato chips every day, because until a few centuries ago no one could get that much salt in their diet no matter how hard they tried. Similarly, for some people, assiduously storing fat whenever it was available used to be advantageous, because fat from hunted game was scarce in some traditional diets. The health risks of obesity that occur when rich food is continuously abundant never came up. But not everyone shares a sensitivity to salt, or a tendency to store fat, and not everyone needs to take precautions against the same pitfalls. Understanding that evolution produces variety helps us understand why one diet won't suit everyone.

This is not to say that many people wouldn't benefit by a drastic revamping of what they eat. We have ample evidence that eating better improves blood pressure, blood sugar levels, and risk of heart disease and diabetes, particularly if we also increase even relatively mild exercise like walking. Increased levels of fiber and less reliance on highly processed foods is almost always beneficial. In one demonstration of this effect, medical researcher Kerin O'Dea and a group of Australian aborigines spent two weeks in the remote outback, on the move most of the day and living off of what they could catch and gather, with their diet comprised of mainly freshwater fish, kangaroo, and wild yams, with the occasional berry and a crocodile included. All of the aborigines had been overweight and several had Type 2 diabetes, but by the end of the trip all showed significant reductions in weight, blood sugar, and heart disease risk. But not even O'Dea emerged from the experience touting kangaroo burgers with a side of yam as the only way to a healthy heart, nor did she claim that it would be realistic to expect modern-day people to mimic this lifestyle permanently.

Those occasional berries probably tasted good, as evolutionarily we expect them to, but there is no reason to feel bad if you would rather have a Snickers than a banana. That doesn't mean you should eat it, but that there is no reason to expect our instincts to be infallible in modern times. How can things that feel so good be so bad for you? It may be the result of evolution, but then, natural selection isn't rational. Presumably, we are, and we should be able to acknowledge our past without letting it dominate the dinner table.

THERE'S NOTHING WRONG WITH YOU

We are growing ever more eager to detect and fix so-called genetic defects in the womb; according to the Genetics and Public Policy Center, a genetic test is available or being developed for more than nine hundred diseases or conditions. At the same time, concern is growing over how much information about potential genetic conditions, particularly incurable ones, people ought to have. Some people envision a world in which information leads to an automatic solution, in which the offending gene is simply plucked out, like a mismatched bead on a string, leaving the remaining pieces in harmony. This fix-it attitude, however, misses something. How did these diseases get into our genes in the first place? Why has natural selection not eliminated them? And why do many strike particular groups of people, like Tay-Sachs in Ashkenazi Jews? An evolutionary perspective helps us see not only our blood and bones but our genetic makeup as the outcome of our longstanding dance with diseases. Some of the diseases may not in fact be true diseases; eliminating all, or even most, of them is as futile a prospect as eliminating our genes themselves.

Cystic fibrosis is the most common fatal inherited disorder among populations of European origin, affecting about one in 2,500 children. The disease causes thickened mucous in the digestive and respiratory tracts, resulting in a buildup of salts and bacteria that can lead to a variety of complications, including difficulty in breathing and an inability to absorb nutrients. Until recently, those affected died young, and even now the expected lifespan is not much over 40. A person with cystic fibrosis must inherit the gene from both parents; people with only one copy are carriers but do not have the disease. The mutation that causes the disease seems to have persisted for at least 52,000 years, which suggests that some advantage kept it from being eliminated by natural selection, particularly since its bearers would have had little chance of reproducing before they died.

It turns out that the mutation may be the footprint of one of those ancient dance steps between parasite and host. Having two copies of the gene is lethal, but having one copy seems to make the bearer less susceptible to cholera and other forms of debilitating diarrhea. The cholera bacteria releases a toxin in the gut that stimulates the intestinal cells to discharge unusually large amounts of salts and fluids, and an untreated victim will die of dehydration after excreting up to 3 or 4 gallons a day. But an experiment with mice that have a single copy of the same genetic mutation showed that they secreted far less fluid than normal mice. Physiologist Sherif Gabriel, who did this research, suggested that human beings with one copy of the gene would have been at an advantage when cholera epidemics struck. Other diseases causing diarrhea, such as *Salmonella* bacterial infection, should also have provided selection for the gene. It isn't clear why Europeans show such high frequencies of the mutation, but it may be that in warmer parts of the world, the increased salt in the sweat of carriers of the gene was a drawback. Unusually salty sweat is a hallmark of those with cystic fibrosis, and until recently, salt may have been at a premium, making the disadvantage of salt loss outweigh the advantage of resisting cholera.

Hundreds, probably thousands, of diseases have similar connections, though in many cases these links are speculative. People with type O blood are more likely to get peptic ulcers. Resistance to malaria is conferred not only by the more widely known sickle cell mutation, but also by hundreds of other subtle blood variations in people from many parts of the world. A "secretor" gene that controls the release of blood-group proteins into the saliva seems to confer protection against bacterial and fungal diseases, but confers susceptibility to viral ones. The gene that seems to cause the autoimmune disease lupus erythematosus is also associated with reduced susceptibility to tuberculosis or perhaps leprosy. Choose your poison. Even genes that influence the body's uptake of vitamin D can affect disease resistance; get osteoporosis, but keep tuberculosis at bay. The threads that go from one gene to another and another are only now starting to be unraveled, and it is very likely that we will discover genetic interactions between traits we never thought could be connected.

In some groups with particular evolutionary histories, genes function in ways they wouldn't if the environment had been different in the past. Lactose, the sugar naturally found in milk and other dairy products, is digested in mammals by an enzyme called lactase. In most populations, the enzyme disappears after weaning, and if people drink milk after they lose the enzyme, digestive problems occur. In some populations, however, most notably Western and central Europeans, lactase production persists, allowing them to consume dairy products with no ill effects. The places where dairy cattle are thought to have originated in Europe also turn out to have the most people with lactase in adults, and the adjustment has occurred within a few thousand years. Do we call lactose intolerance a disease? Presumably not; it is one of the many forms of variation that influence how the body functions. So what about other kinds of genetic variation, for tolerance of foods or sensitivity to light, or myriad other quirks—do we attempt to get rid of them, simply because we never made the right connection between the variant and a disease?

Genes don't just sit there, like those metaphorical beads, neatly coding for one trait and one trait only—a gene for eye color, a gene for leg length, for intelligence, for a disease. There are no "sickness genes," and though many mutations can be quite harmful, a gene that did nothing but render its bearer sick, or susceptible to becoming sick, would never survive. This is not to say that everything from cystic fibrosis to mental retardation has a purpose, or that we should not try to correct the defects we see. For diseases like cystic fibrosis, where the ill effects are seen only when two copies of the mutation are passed on, there is particular scope for genetic correction. Some of the selective forces that have left their mark on our genomes disappeared thousands of years ago, and for these too manipulation may be feasible. Hypertension has been suggested to be higher in people who had a high risk of water deprivation in their evolutionary history, making salt retention critical; such a threat is far less likely now. Sickle cell anemia occurs in people with two copies of a gene that confers resistance to malaria if it is present in just one copy; it would be better to prevent malaria in other ways than to rely on the genetic resistance with its collateral damage of anemia. But an appreciation for our long relationship with disease-causing organisms— and thus for Darwinian medicine—should make us careful about blithely condemning all genetic peculiarities as meaningless, or worse, requiring medical removal. After all, what new disease challenges, requiring us to meet them with the raw material of our variable genes, might arise next?

CHAPTER  2

FRIENDLY WORMS AND THE PRICE OF VICTORY

Imagine living in a hunter-gatherer society in Africa several thousand years ago. What kind of illnesses would you be most likely to suffer from? Because your group is relatively small, on the order of several dozen to a few hundred inhabitants, you need not worry about diseases such as measles (which require a population size of several thousand to remain viable) but you, your family, and your friends harbor several different kinds of intestinal worms, and malaria is common, particularly among children. Depending on where your group lives, schistosomiasis, a liver fluke disease, may also be abundant. Blood-tinged urine may be so common among young boys that it is viewed as a rite of passage similar to menstruation in girls. No one wears shoes, facilitating the transfer of worm eggs, bacteria, and plain ordinary dirt from bare feet to hands to nose and mouth, where they wend their way to the intestinal tract.

Now fast-forward yourself to eighteenth-century Paris. Some of the worms are still there, and in addition you and your neighbors have seen cholera sweep through the city, carried in the communal water sources necessary to supply the growing population. Smallpox is also a scourge, and a great many babies die of viral and bacterial infections before they are toddlers. The domestication of animals like cows, pigs, and sheep has brought their diseases to humans. Sedentary living has also exacerbated the transmission of disease, as communities grow up around their own waste. Sewage disposal is primitive at best, so if one person has diarrhea due to *Giardia,* a one-celled parasite that lives in contaminated water, chances are that those around him get it too. In both societies, if a cut or wound becomes infected with bacteria, the resulting "blood poisoning" is often fatal, and such secondary infections of injuries sustained in battle are more likely to kill a soldier than the saber cut or musket shell itself.

Although plenty of diseases still threaten us, most people in today's western industrialized countries do not have to worry about any of the diseases so prevalent in earlier times. Smallpox has been completely eradicated; cholera and schistosomiasis can be controlled by effective sanitation measures; vaccination prevents measles, diphtheria, and a variety of other illnesses; and drugs developed in the mid-twentieth century cure most bacterial infections, though the wounds of war are still with us. Our homes and bodies are so clean, so free of dirt and microorganisms, that they would be nearly unrecognizable to our forbearers. No one would want to trade our lives for theirs, at least in terms of overall health and risk of dying from infectious diseases.

And yet some modern diseases would perplex the African tribesman or the Parisian of centuries past. Asthma, for example, would have been virtually unknown, as would hay fever and allergic sensitivities to substances ranging from peanuts to pollen to pet hair. Cave dwellers almost certainly did not spend the spring sneezing and blowing their noses on animal skins. They, and most other people from before the twentieth century, as well as those from less developed nations today, lacked an entire suite of illnesses that seem to result from the immune system backfiring on itself, reacting to environmental inhabitants like pollen as if they were deadly invaders instead of benign foreign objects. Scientists now identify the so-called "diseases of the advantaged," including not only allergies and asthma but also inflammatory bowel diseases such as ulcerative colitis and Crohn's disease, all of which have skyrocketed in their frequency in developed nations. And the number of cases continues to climb; asthma rates increased 75 percent in the United States between 1980 and 1994, but remained essentially unchanged in less developed parts of the world. Eczema and other forms of allergic skin reactions are also on the rise, along with food sensitivities. What's more, the various ailments are connected. Physicians and parents alike talk about the developmental Allergic March, in which an infant plagued with eczema turns into a child with asthma who then becomes an adolescent and adult with hay fever.

Where has this epidemic come from? There is no shortage of theories, but some of the obvious ones, like the prevalence of pollution, can be rejected, at least as the sole explanation; asthma levels are just as bad in bucolic Rochester, Minnesota, as they are in grimy Chicago. Environmental agents such as soot particles in the air can certainly trigger an asthma attack once a person has developed the disease, but they do not seem to be the initial cause of the illness. Also, asthma and many allergic diseases are virtually absent in parts of Africa, China, and other less developed nations, even though pesticide use and pollution may be high. Genetic differences, though they may play a role, again cannot be the sole explanation because the rise in these diseases was too precipitous to be caused by the slow changes in gene frequencies in populations. The diseases are probably not due to changes in diet, because the diets of afflicted individuals seem to share few or no foods, nor did diets change uniformly and rapidly in the various countries where these diseases occur.

What is more, a curious relationship exists between the number of children in a family and the likelihood that one or more of them will develop such a disease: More siblings means less risk of asthma, regardless of whether the family is in pristine

Montana or polluted Los Angeles. Having older siblings is particularly likely to be associated with decreased incidence of asthma and allergies, as is—perhaps counterintuitively—the presence of a pet in the household when a child is born. Children in rural environments are also less affected by the diseases, whether in industrialized or less developed countries, and those who recall spending time in a barn and drinking unpasteurized milk are particularly untroubled by allergies. None of these links are absolute, and plenty of only children without a pet to their name still escape asthma, while many from homes filled with cats, dogs, and brothers and sisters still suffer, but the relationship is marked enough to have made health-care workers wonder about its source.

THE HYGIENE HYPOTHESIS

Dr. Erika von Mutius, a German medical researcher, compared the incidence of asthma and allergies in East and West Germany just after their reunification in 1989. She theorized, reasonably enough, that the more polluted environment, lower standard of living, and worse health care of East Germany would cause a higher rate of the diseases there than in the relatively cleaner West. As you can probably guess by now, however, the opposite turned out to be true, with children from East Germany being much less likely to suffer from allergies and asthma. She, along with several other scientists who had noticed similar phenomena, proposed the hygiene hypothesis to explain the results. The hygiene hypothesis suggests that diseases such as asthma and allergies arise from an environment that is too clean, so that the normal stimulation of the immune system during infancy and early childhood is missing, impairing its ability to respond normally to actual pathogens but ignore harmless entities like pollen.

In other words, day-to-day exposure to bacteria, viruses, and other microorganisms provides a model for appropriate defense behavior, so that the immune system doesn't cry wolf every time an errant dust particle comes its way.

Again, recall that all living things evolved with parasites and disease, and our bodies grew accustomed to dealing with them. Remove the pathogens, and it is as if the immune system casts about for something to do. This is not to say that being sick is good for you, but it is starting to seem that never or rarely being sick can be bad for you.

Evidence is accumulating that contracting all manner of diseases early in life protects against developing some forms of allergies and similar autoimmune disorders. Scientists, including Robert Desowitz, the author of several wonderful books on parasites and people, have noted since the mid-twentieth century that people in areas where malaria was rampant had far fewer autoimmune disorders. Dr. von Mutius and her colleagues collected information from twenty-three countries in Europe, North America, and the Pacific on the rates of positive tuberculosis tests, which indicate exposure to the bacteria that cause the disease, and found that the incidence of asthma and rhinitis-type allergies (hay fever) was lower where the rate of tuberculosis notification was higher. If Italian military recruits (a good captive audience, and presumably a reasonably random sample of young men) had signs of previous exposure to hepatitis, or to a relatively benign parasite called *Toxoplasma,* or to *Helicobacter pylori,* the bacteria that cause stomach ulcers, they were less likely to suffer from asthma or hay fever. An examination of 520 African children in Gabon showed that they were thirty-two times less likely to have a skin sensitivity to house dust mites, a common allergen in the west, if they also showed evidence of schistosomiasis in their urine. Even the lowly sniffles from a cold are associated with reduced incidence of wheezing caused by asthma, or, as pediatric allergists Andrew Liu and James Murphy put it, "the cure *is* the common cold," at least for children under age seven with allergies.

Several studies have zeroed in on the protective effects conferred by exposure to endotoxin, a component of the cell walls of a large group of different kinds of bacteria. Endotoxin occurs naturally in the environment, and tends to be particularly high in places like farmyards where large animals such as cows, sheep, and horses live, probably because minuscule bits of the animals' fecal material are borne in the air and soil. Endotoxin is also present in varying levels in dust and dirt both indoors and out, and is easily inhaled or swallowed. When this happens, several molecules in the immune system are stimulated, and these in turn signal a cascade of other responses. Charlotte Braun-Fahrlander and a group of European medical researchers, including von Mutius, measured the levels of endotoxin in bedding used by 812 children between the ages of six and thirteen in farming and nonfarming households from rural areas of Germany, Austria, and Switzerland. They also examined the immune system responses of the children and the reported symptoms of asthma and allergies. As you might expect, the higher the level of endotoxin in a child's mattress, the lower the likelihood of that child sneezing, wheezing, and sniffling his or her way through to adolescence.

These results may help to explain the association between animals, including pets, and decreased asthma incidence. The correlation between number of children in a family and rates of allergy probably occurs because siblings are excellent vectors of bacteria, viruses, and other pathogens, and is also consistent with the finding that time spent in day-care centers can accomplish much the same thing as having a lot of brothers and sisters to lavish germs upon each other. Parents can bemoan the frequency of colds and other infections during the first year a child is in day care, but the sicknesses may help to protect a child against allergies. High levels of asthma among inner-city children may occur not because they are living in dirty environments, but because they are dirty in the wrong way, with lead and soot rather than cow droppings; perhaps we should encourage city dwellers to take up dairy farming.

Food allergies are less often considered in light of the hygiene hypothesis, but that may change given the results of a remarkable study by a group of Norwegian researchers who looked for a connection between childhood sensitivities to eggs, fish, or nuts and whether babies had been born vaginally or via cesarean section. Childbirth is a rather messy process, and along with an introduction to air, light, and exclamations of joy, a newborn receives a hearty dose of bacteria to colonize his or her intestinal tract during the passage through the birth canal. Babies removed surgically from their mothers miss out on this germ-laden opportunity, and the scientists hypothesized that the difference in natural bacterial inhabitants might influence the way the infants responded to known triggers of allergies in food, much the same way that being exposed to endotoxin or other normal microbes regulates other allergic responses.

Although most historians are probably blissfully unaware of it, one of the marked changes in our lives since the nineteenth century is in the number and kind of microorganisms we carry in our guts. We harbor hundreds of species of bacteria in our gastrointestinal systems when we are healthy, an internal forest of biodiversity. The internal flora in western babies at the start of the 1900s was similar to that of babies in less developed nations today, but different from that of modern western infants, who now harbor various microbes acquired in hospitals. Babies delivered by cesarean section acquire their intestinal menagerie more slowly than vaginally delivered children. This is not automatically a bad thing, but once again, it is not the way we evolved. The scientists surveyed 2,803 children and their mothers, obtaining information on any food allergies of the mother, the existence of pregnancy complications, the birth weight, and use of antibiotics by either the mother or her baby. They looked for food allergies at age two and a half, both by asking the mothers about their children's reaction to the foods and by performing sensitivity tests for egg protein in a doctor's office.

When a mother herself was allergic, her child was seven times more likely to have a perceived allergy to the three substances if she had experienced a cesarean section, even after controlling for the other factors like size of the baby and whether delivery had been premature. By itself, cesarean delivery was associated with a small increase in the likelihood of food allergies, but this was not different from what might have occurred by chance. Only 13 percent of the women in the study had given birth via cesarean section, however, and it is possible that a larger sample size, reflecting the greater than 20 percent cesarean birth rate in the United States, for example, might have shown a stronger association. Breast-feeding did not have an effect on allergy in their study (though it has been shown to be beneficial in many other ways), nor did the use of antibiotics in either the mother or her child. The authors of the study recognized that cesarean delivery is often unavoidable, but they point out that if the procedure is elective and the mother allergic, she might consider this potential increase in her child's likelihood for developing food allergies.

It is important to provide an additional cautionary note here, and to realize that these associations do not automatically mean that if some is good, more is better; extremely high levels of endotoxin are associated with increased wheezing in children, and once a child is past a certain age, endotoxin and other environmental impurities can actually worsen allergic responses. Then too, as I mentioned earlier, genetic and other factors, such as smoking, also play a role in the development of diseases like asthma. Asthma itself can be tricky to diagnose, because it is a collection of symptoms that are sometimes evaluated subjectively. Not all studies support the hygiene hypothesis; some find no link between asthma rates and hygiene. And as the author of an editorial in the *New England Journal of Medicine* pointed out, "Eating dirt or moving to a farm are at best theoretical rather than practical clinical recommendations for the prevention of asthma." Nonetheless, support is growing for the idea that parasites and illness are a natural part of life, and that trying to eliminate them is both futile and likely to cause unforeseen problems.

THE WELL-TEMPERED T CELL

Exactly what is it about early stimulation by bacteria, viruses, or other parasites that keeps the immune system calm in the face of harmless entities like pollen or one's own intestinal cells? Obviously the analogy of immune system cells being like bored unemployed workers that make mischief on the rest of the body is just that, an analogy, and even the most anthropomorphic among us stops short of assigning personality traits to bone-marrow products. The more accurate answer seems to lie in a characteristic of the immune system. Part of our response to foreign invaders of the body is mediated by a kind of white blood cell called a T cell. The T cells come in a variety of types, including killer T cells and helper T cells. The helper T cells in turn are also divided into two types, called Th-1 and Th-2. The Th-1 and Th-2 responses are responsible for protection against different things, with the former concerned with bacterial and viral diseases and the latter with infections by worms and other large parasites. Each type of helper T cell produces a different set of chemical messengers used to regulate inflammatory responses like tissue swelling and the production of mucus. These chemicals interact with each other and keep the entire system in balance.

In countries with scrupulous hygiene, where children are vaccinated and antibiotics are widely administered, the low level of Th-1 stimulation results in an increase in the Th-2 response. These Th-2 responses trigger an exaggerated mucus production and contraction of muscles in the airways, which can in turn cause allergic diseases and asthma. In countries where bacteria, worms, and other pathogens are abundant but vaccination and antibiotic levels are low, the Th-2 responses are activated, but they are regulated by repeated cycles of infection and inflammation, with the inflammation countered by natural antiallergic reactions. Thus they rarely escalate out of control as much as Th-2 responses in people from the industrialized areas. The immune systems of people from less developed countries still respond physiologically to allergens like pollen or house dust mites, but the people do not go on to develop a disease. It is as if the Th-2 arm learns to recognize an innocuous but foreign substance for what it is, and has a blasé "been there, done that" reaction to it, rather than spiraling into a panicky cycle of swollen tissue and dripping glands.

NOBODY LIKES ME, EVERYBODY HATES ME

Never mind the suggestion, facetiously posed and hastily dismissed, that we should be eating dirt. According to Joel Weinstock, professor and chief of the Division of Gastroenterology/ Hepatology at Tufts New England Medical Center in Boston, we should actually be eating worms, and he is serious. Well, not all of us—just the growing number of us who suffer from Crohn's disease, an inflammatory bowel disorder that causes abdominal pain, diarrhea, and complications ranging from intestinal blockage to sores inside the digestive tract. Fatigue and fever are common, and bleeding within the intestines can cause anemia in some Crohn's sufferers. In younger patients, inadequate absorption of nutrients because of a chronically inflamed digestive system can hinder growth and cause other disorders. The disease comes and goes with periods of flare-ups and remission, and in severe cases large portions of the intestine need to be removed surgically. Treatments include a variety of antibiotics and powerful immunosuppressive drugs such as corticosteroids, many of which have severe side effects. In about 20 percent of patients, a close relative, perhaps a brother or sister, also has the disease, suggesting a genetic link.

The cause of Crohn's disease is unknown, but several years ago Weinstock and some of his colleagues began to wonder why it had sprung to medical awareness so recently. The disease was first recognized among Jews living in New York City in the 1930s, and then began to be diagnosed in non-Jewish Caucasians, followed by people of all ethnic backgrounds. It also became clear that it showed a gradient from the northern states, where it was more common, to the southern states, where it was rarer. The same geographical trend occurred in Europe, with Scandinavians and Britons suffering more than those from sunny Mediterranean countries. Crohn's is also rare in most parts of Africa, Asia, and South America. Military veterans who were prisoners of war or served in the tropics are less likely to have the disease. And the frequency of inflammatory bowel disease has kept on climbing; Weinstock says, "Once this was a disease that people spoke of in terms of one in 50,000, then one in 10,000, then one in 5,000, and now one in 250. Crohn's disease is a disease that is common, not rare."

So what's changed? Obviously, many things, but Weinstock's group wondered if the problem wasn't so much something people had acquired, like environmental toxins or a bad gene, but something they had lost. Specifically, they had lost the parasites that usually inhabit people in agricultural areas, or places where contact with soil is common. People had begun to wear shoes, keeping their toes safe from broken glass, hot asphalt—and worm eggs. Living in northern climates means that the worms that evolved to spread in warm tropical soil were thwarted. And many Jewish children did not eat the pork that carried one of the most common intestinal parasites of pigs and humans, which could explain the apparent link between Crohn's and being Jewish, at least initially. In the 1940s, one out of six Americans showed signs of exposure to *Trichinella,* the pork-borne parasite, but by the 1960s this proportion was less than 5 percent, with only half of 1 percent with indications of a recent infection. Furthermore, starting in the twentieth century, children and adults were being given drugs to rid themselves of worms, something new in our evolutionary history. According to Weinstock, "If you simply follow inflammatory bowel disease around the world, wherever there is deworming you are seeing a rapid rise in inflammatory bowel disease frequency."

The worms are thought to stimulate that Th-2 response in the immune system, which then helps regulate Th-1 activity. A malfunctioning Th-1 response is characteristic of Crohn's disease, and so Weinstock made the bold suggestion that worms could cure Crohn's, or at least help control it. Workers in his laboratory first tried infecting mice with either worms or the fluke that causes schistosomiasis, and the results were encouraging:

The mice were protected against developing an intestinal inflammation similar to Crohn's disease, and they manufactured the "right" sort of chemicals to regulate their Th-1 immune responses. Genes also play a role in the inflammation by controlling the production of these regulatory chemicals, but this does not mean that Crohn's is an inherited disease; no single gene or set of genes has been shown to determine whether an animal or person gets one of the inflammatory bowel disorders.

Then the scientists started contemplating worm therapy for humans. Obviously, even though we have evolved with intestinal parasites, they are still parasites, and they can still harm their hosts, particularly if the host is malnourished or has other health problems. Regressing to the era when most people had worms themselves is not a viable solution to the problem. Besides, where would one obtain the worms to use for infection? It is hardly practical—or ethical—to keep a stable of worm-producing people as a pharmaceutical source, and it is important to regulate the source of the immune stimulant.

So Weinstock and his colleagues decided to try a whipworm that is found in pigs. Whipworms are very common parasites; the human whipworm has been found in mummified remains and fossilized human feces from over 10,000 years ago, and people all over the world used to harbor infections from it. These days virtually no one in industrialized countries has whipworm, but a close relative of the kind that infects humans occurs in pigs. The pig whipworm, called *Trichuris suis,* does not cause human illness because it cannot establish itself in people. Most parasites are highly specialized, and being in a new host is like trying to live on a different planet; the surroundings are so different that even if you can manage to survive, you almost certainly will not be able to reproduce. In a person, the pig whipworm will stimulate the Th-2 response, but after a few weeks it will die and be passed out of the body. The scientists could get pig whipworm eggs from pigs that are reared in a special environment free of other pathogens, so that the worm eggs are sterile. They were reasonably sure that the eggs would be tolerated well because pig farmers have been exposed to the worms for hundreds if not thousands of years, and do not become ill. As a final safeguard, drugs that completely eliminate the infection are available in case someone had a bad reaction, or second thoughts.

The next hurdle was trying the treatment in human subjects. Weinstock hastens to point out that patients are never asked to swallow, à la *Fear Factor,* live wriggling worms. They are given a solution of the nearly microscopic pig whipworm eggs in Gatorade (there is an irony there, but I am not sure exactly what it is), and then they wait for the worms to do their stuff. The worms settle in the right side of the colon, where they perch looking unsettlingly perky in a photo from one of Weinstock's papers, but they cannot multiply or leave the intestine for another part of the body. In a preliminary study, four pioneers with advanced Crohn's disease and three with ulcerative colitis, a related inflammatory bowel disease, were each given 2,500 worm eggs. They all knew they were receiving the treatment, rather than a placebo, but much of the scientists' initial interest was in monitoring side effects of the worms, ensuring that they did no harm. Much to their satisfaction, three of the Crohn's patients achieved remission, meaning their symptoms essentially disappeared, and the fourth showed a marked improvement, as did all of the ulcerative colitis patients. Not one showed a single side effect of treatment, unlike the situation with conventional drugs, which can cause hair loss, swelling, nausea, and a host of other symptoms.

The effects of the first dose lasted two to five months; repeated doses every three weeks or so seem to keep the immune system regulated appropriately. A second study was completed in

2004, using twenty-nine patients with ulcerative colitis. This time, the study was double blind, meaning that half the patients received an inert placebo and half the real thing, and neither patients nor those administering the dose knew who got which treatment. After twelve weeks, those who had received the worms were switched to the placebo and vice versa. About half the people ingesting worms showed significant improvement, compared to only 15 percent of those receiving the placebo. In another test of the therapy with Crohn's patients, nearly three-fourths of worm recipients showed remission after six months of taking the treatment every three weeks. Rates of improvement might be higher in larger-scale or longer-lasting tests, since only those patients who had not responded to standard therapy were used in the first sets of patients.

Of course, there is the ick factor to consider. Or, as Canadian medical researchers M. M. Hunter and D. M. McKay put it, "the individual patient's perception of helminth [worm] therapy and their adjustment to the concept would be a pivotal decision branch in any logarithm of potential treatment approaches to colitis." In other words, if people are too grossed out, they won't be willing to undergo the therapy. But Weinstock told me that people "have little resistance to such treatment. Really." After all, he pointed out, these patients have already been on drugs with potentially serious side effects, taking over 20 pills a day. The prospect of relief from their symptoms seems to outweigh any squeamishness. Then, too, he is clearly a worm enthusiast; in a BBC interview he said, "People have what I consider an irrational fear of worms." A book chapter from his laboratory gleefully lists seventeen kinds of worms commonly found in humans. And in virtually all of his interviews and articles, he manages to work in what seems to be his signature statement: "Half the weight of your stool is living bacteria. What are a few worms more or less?"

What indeed. This upbeat attitude notwithstanding, the Tufts researchers are working with a German firm to market a drug that will be a bit more palatable than simply giving people worms. And eventually, the goal should be to figure out how to prevent diseases like Crohn's, so that the immune system does not go awry in the first place, and do it without giving people actual infections with parasites. A dose or two of the worms themselves is not a permanent cure for inflammatory bowel disease, which makes sense given that chronic, low-level infections with a variety of parasites are part of our evolutionary history; a single dose of any agent, no matter how helpful in regulating immune response, cannot replicate a lifetime of exposure.

WHO IS FOR WORMS

So if harboring some worms can be good for you, what does that say about our efforts to control parasites in developing countries, where they are still prevalent, particularly among children? Previous goals have always been to eradicate worms completely. But the World Health Organization recently started a parasite-control effort that may well be more successful than more ambitious ones, precisely because it sits at the balance of too much and too little treatment. Although small doses of worms are manageable in well-nourished hosts, children in these countries often have nutritional deficiencies and other health problems, and the effects of the worms are more serious, causing learning difficulties and stunted growth. A recently launched effort to control these diseases, the Partners for Parasite Control program, a cooperative effort by United Nations agencies, nonprofit groups, and research institutions, has as its goal the treatment by the year 2010 of at least 75 percent of all school-aged children who are at risk of the detrimental effects of blood flukes and soil-transmitted worms. Previous efforts to control such infections have met with mixed success, to say the least, with the proportions of infected children remaining high despite repeated treatments. The new program is optimistic about its chances because of three key changes in its approach. First, drugs to treat worm infections have dramatically decreased in cost in the last several years as patents by drug companies have expired; the annual cost of treating a child for all of the common parasites is now less than 50 cents, within the reach of poor countries. Second, the program seeks to piggyback onto existing networks like school systems, training nonmedical people such as teachers to administer the pills to their pupils. Earlier programs were externally funded, so treatment efforts collapsed when the program's support was withdrawn.

It is the third difference in approach, however, that is the most significant from my perspective. Instead of attempting to reduce the prevalence of infection, represented by the number of people who have any worms at all, the Partners for Parasite Control program accepts the inevitable reinfection that accompanies poor sanitation. They recognize that clean water and proper waste disposal are laudable, but in the meantime, the Partners argue, the goal is to reduce the number of worms a person has and the intensity of infections within individuals, so that continual treatment keeps the worm burden at a very low level. Not absent, just minimal, which drastically decreases the parasites' effects. The point is not to ban worms, but to live with them. Past efforts to control parasitic infections took a much more unyielding line, insisting on completely curing individuals, and generally failing to do so.

In an ideal world, of course, everyone would have access to clean water and good sewage management, and the complete eradication of worms would be a more achievable goal, though perhaps not an altogether desirable one. But the Partners' more modest aim of control of parasite numbers contains the recognition that we, like all other species, live with parasites. Without intending to, WHO is taking an evolutionary approach. Intestinal worms are a part of our heritage, and, it might be argued, only became such a dramatic public health problem when burgeoning populations and overcrowding led to an increase in transmission.

THE SEA WITHIN US

Our bodies did not evolve with only the odd worm or two; they also have millions of other inhabitants, happily infesting every nook and cranny, every fluid and tissue. Never mind what goes out, à la Weinstock's cheerful assertion about stool composition; currently inhabiting your intestinal tract are between 500 and 1,000 different kinds of microbes, like a teeming rain forest of diversity. While not as colorful or as noisy as parrots and monkeys, they perform different tasks and communicate with each other, and disturbing their balance can be like clear-cutting the Amazon jungle, leaving a bleak devastation behind. Or to make the analogy a little more personal, think of your gastrointestinal system as a complex border clearinghouse for everything that comes into your body. Someone has to manage immigration and customs, barring dangerous visitors and encouraging those with something to offer. Your own cells can only do this to a limited extent, and rely instead on outsourced labor to protect the internal organs from harm while allowing nutrients to pass through. This relationship is a very ancient one, so ancient that many biologists believe that the different components of our cells—the nucleus with its genetic material, the mitochondria that serve as the energy sources—originated as separate beings, bacteria that became so intimate with their hosts they were enfolded into the cell itself and became a part of us. The bacteria in our guts have not taken quite such extraordinary liberties, but they are still blushingly close, more in tune with our needs than the most intuitive lover.

The line between friend and foe can be a precarious one. Some gut bacteria are innocuous in small numbers but potentially troublesome if they multiply too quickly, while others need to exist in the appropriate balance with the rest of the menagerie to do any good. Because of this blurry line, these generally nonpathogenic bacteria are called commensals, the same term that refers to other animals that live together without harming each other.

We do not automatically possess this internal anthology of microbes; it is assembled when we are born, and from our food and environment. In this regard we act like termites, albeit in a rather more primitive way. Eating wood is a difficult task, because the cellulose it is composed of has rigid cell walls that animals cannot digest by themselves. Termites harbor microorganisms in their intestines that do the digestion for them, and after a young termite hatches from the egg, it must seek out and feed on liquid provided by an older individual in the nest. Our own method is more haphazard—we inhale and touch and lick up the microorganisms that come our way from birth onward, gradually accumulating our lifelong hitchhikers. If mice are reared in a germ-free environment, one without this internal stew, they must consume about 30 percent more calories to maintain the same body weight as mice kept in a more conventional, bacteria-ridden habitat. Our beneficial relationship with these microbes has a long evolutionary history; some types of intestinal bacteria also seem to have an encouraging effect on various immune cells and tissues, suggesting their longstanding familiarity with our bodies.

Just as our modern lifestyles have discouraged worms, they have dramatically changed the types and number of bacteria in our bodies. Two factors are responsible for this internal biodiversity crisis: hygiene and diet. Our antibacterial crusade has gotten way out of control, seeing bacteria as the enemy, to be vanquished at all costs, rather than an essential companion. As for diet, people used to preserve foods by drying or fermenting them, then acquiring a hefty dose of bacteria when the food was eaten. Processed foods contain fewer bacteria, as do products from animals that have been given antibiotics to promote growth and prevent disease, as most farm animals have. This shift has been more subtle than the removal of large parasites from our lives. Nevertheless, like a town that has gone from having meadowlarks, finches, and a dozen different kinds of warblers before urbanization to a monotony of human-introduced starlings and pigeons afterwards, the species composition of our guts has changed.

Recent studies of some common gastrointestinal disturbances, including traveler's diarrhea, the disorder that often accompanies trips to tropical or less industrialized places, suggest that administering a dose of "good" bacteria can prevent or at least ameliorate the problems. The most common bacteria used are *Lactobacillus*, found in yogurt and other cultured milk products, and *Bifidobacterium*, and they are called probiotics, in contrast to the antibiotics that act to kill bacteria. Doses of these microbes seem to stabilize the other gut flora, and can also help the gut resist a common pathogenic virus called rotavirus. Sometimes the probiotic is accompanied by a molecule called a prebiotic, which helps promote the growth of the normal good bacteria.

The rationale for probiotic therapy is similar to that behind the hygiene hypothesis, except that here, instead of people needing to be exposed to the "natural" kinds of diseases during childhood, they need to be exposed to the appropriate colonies of microorganisms in their gastrointestinal tract. Probiotics have also been suggested as treatments for diseases as diverse as rheumatoid arthritis, depression, and urinary tract infections, and one type of *Lactobacillus* may reduce the recurrence rate of bladder cancer.

Some of the best evidence for probiotic therapy comes from its use in reducing eczema in babies and children. Eczema, more properly called atopic dermatitis, is an itchy skin rash that occurs spontaneously but can also be triggered by heat or cold or contact with skin irritants like certain detergents or soaps. It is associated with the other diseases in the Allergic March, as I discussed above, and tends to run in families. A group of Finnish researchers gave *Lactobacillus* to pregnant women who had a family history of asthma, seasonal allergies, or eczema, and also supplemented their infants with the bacteria until six months of age. They used a double-blind study with a placebo, and had physicians confirm the presence or absence of the skin condition. Half as many children who had received the probiotics developed eczema, a striking reduction in the disease. A follow-up study four years after the children were born showed a persistent effect of the treatment, suggesting that the changes in the bacterial inhabitants of the supplemented youngsters were long lasting.

A different kind of bacterial supplement, *Bifidobacterium,* was used to treat patients with ulcerative colitis, one of the inflammatory bowel diseases in the same class as Crohn's disease. Here the bacteria were thought to help regulate immune system molecules similar to those thought to have been modified with the Weinstock worm therapy. The bacteria were combined with one of the prebiotics to help them survive in the gut, and although the study was small, treated patients showed a marked improvement in their condition, and research is continuing.

Perhaps not surprisingly, ingesting friendly bacteria has seemed much more appealing to the general public than the idea of consuming worms. This has led to a proliferation of nutritional supplements containing bacteria as well as some medical claims from alternative sources that are, to use a technical term, really out there. Again, caution is warranted, as some bacterial supplements were found not to contain the substances reported on the label. It is too early to suggest that the average person start swallowing bacteria, much less worms, but it is likely that some kind of probiotic therapy will become an accepted part of the medical arsenal in the near future. In the meantime, eating yogurt with live cultures may be an acceptable stopgap.

CLEANLINESS IS NEXT TO . . . SICKLINESS?

So how have we come to this state, where we need to inoculate ourselves with bacteria to be healthy and are getting sick because we lack the normal exposure to parasites as children? Part of the answer, of course, is that these are side effects of a dramatic reduction in childhood mortality from disease. No one is suggesting we go back to some imagined nirvana when everyone had intestinal worms and our food teemed with bacteria, good and bad. But we also seem to have fallen prey to an obsession with germs and cleanliness, an obsession that goes so far beyond the prevention of catastrophic diseases like tuberculosis or cholera that it creates catastrophes of its own.

Between 1992 and 1998, about seven hundred new antibacterial products came on the market in the United States. That's a lot of hand sanitizers, laundry detergents, computer keyboard covers, countertops, and room sprays. People are urged to wash their hands with special soaps, to microwave their kitchen sponges and disinfect their teeth. And although Howard

Hughes's excessive hand washing was a sign of his obsessive-compulsive disorder, didn't more than a few of us secretly find his germ phobia perfectly natural?

The idea that invisible yet ubiquitous beings like bacteria and other microorganisms can be dangerous is enough to give even the most placid a twinge of paranoia. But many of the efforts to make our environment and ourselves cleaner are at best ineffective and at worst more harmful than the microbes they aim to destroy. Take, for example, your doormat. In February 2005 the *New York Times* reported on Debbie Estis Greenspan, who suffered from allergies and hoped her daughter Haley would be spared the same fate. After seeing Haley crawling on the carpet and discovering that the floor could harbor millions of bacteria ("from fecal matter," the *Times* shudders, clearly unaware of Weinstock's cheery factoid), Greenspan hit on the idea of Dr. Doormat, which contains a chemical that kills microorganisms before they are tracked into the house. Now Haley can loll on the rug without fear of ingesting germs tracked in on people's shoes.

Anyone who read the first part of this chapter probably realizes that this is exactly the wrong approach for reducing the likelihood that children will suffer from asthma and allergies. No one wants Haley to be licking *Salmonella* off the floor, but a hearty dose of endotoxin would probably do her a world of good.

At some level, we can hardly blame the Greenspans. It is a case of technology outstripping necessity. If you live in a hut, or even a log cabin, and you or your neighbors keep livestock, and your water comes from a local well or the river, working under the rubric of "keep dirt out of the house and wash away grime as much as possible" makes sense, and you will simply never be able to kill so many microorganisms that your immune system goes awry. Now, however, we possess the capability of sterilizing our environment to an extent our ancestors could never have imagined, and so our laudable impulses have run amok because the tools to make those impulses reality have become more powerful. Just as the rule to seek out sweet, fatty foods has backfired in a world of Ben & Jerry's and Big Macs, so too has the rule of keeping ourselves away from all potentially infectious agents.

The frightening consequences of our overzealous cleansing include resistance of common bacterial infections, like staphylococcus in hospitals, to most or all of the antibiotics we use to control them. Sometimes, however, even if it does not have such deadly implications, our fear is out-and-out groundless. Periodically magazines and newspapers carry admonitions about our toothbrushes being breeding grounds for nasty microorganisms, and that we should be vigilant about changing them, particularly if we have been sick, to avoid reinfecting ourselves. Numerous consumer health groups advise getting a new toothbrush as often as every two weeks, and a purveyor of an antibacterial toothbrush purifier warns about keeping the toothbrush in a noxious, germ-laden "outhouse," as it calls a modern bathroom. Its Web site abounds with horrific descriptions of the germs that you could be introducing into your mouth with every swipe of your gums. "Bacteria thrive on toothbrushes," one article warns.

People, let's think about this. If you are sick, say with a cold, you certainly have the offending virus in cells in your mouth, as well as on various other body surfaces. To combat the illness, your immune system produces antibodies that render the virus harmless, and contain a memory of that particular culprit, so that if you are ever infected again, similar antibodies rise up immediately and vanquish the invader (I know, I railed against these war images earlier, but they do get the point across). With a cold, that initial immune response takes about a week. This means that you cannot reinfect yourself with the same pathogen, using your toothbrush or anything else, which is a good thing, since if you could, the virus or other microbe that started out in your mouth to begin with would perpetuate an endless cycle of disease. If you develop any other infections, they could enter easily through a variety of portals, not requiring a solicitous introduction via your toothbrush bristles. What would be the point of killing the handful of bacterial cells on your toothbrush, when their source, the mother ship of microorganisms, is inside your body to begin with?

Sharing a toothbrush with someone else changes all these rules, needless to say, as does using a toothbrush that creates tiny cuts in your mouth, or dunking your brush into scummy pond water, since these present new avenues for infection. But significantly, the American Dental Association calmly suggests changing toothbrushes every three or four months, as they become frayed, and makes no recommendations about what to do after an illness, dental hygiene-wise. And if bacteria thrive on a toothbrush, they do so in a pale imitation of what they are doing all the time inside your mouth.

Undaunted, the products keep on coming. One of the latest was a blister pack of lip gloss in individual units, because "Dipping germ-coated fingers into lipcolor could cause last week's cold to reappear." Again, if your fingers could give yourself the same cold you had last week, you'd never get well. Of course, the packs of fifteen individual portions cost ten dollars, so someone is doing all right.

My own personal favorite of the germ-phobic ads is one for a line of antimicrobial products for the office, including computer keyboards and desks. The promotional material states, "Researchers found that the average desk had 400 times more bacteria than the average toilet seat." I had several reactions to this. First, my inner statistician wondered how one determines the "average" desk, not to mention toilet seat (public restroom? House with toddlers? Fraternity?). And then, being a glass-half-full kind of person, I found it pretty impressive that toilets are that clean. Maybe we should just rest on our laurels, so to speak, and figure that if a toilet has that few bacteria, the rest of the world is probably innocuous enough. Reading the ad a little more closely reveals that the products are actually designed not to prevent disease but to keep the wood and other material from getting stains and mold caused by microbes. Presumably the manufacturer relies on the yuck factor to draw customers.

Even carefully controlled tests of many antibacterial products fail to find a benefit to their use. One study took 124 households and had half use antibacterial soaps, laundry detergents, and cleansers, while the other half used ordinary products. All of the households had at least one child in preschool, and all were monitored for 48 weeks for symptoms of colds, flu, and food poisoning. At the end of the test period, the two groups had the same incidence of illness. Another group of researchers examined whether those reassuring-looking gloves worn by food service workers really prevent bacterial transmission, by sampling the microorganisms on tortillas purchased from gloved or bare-handed restaurant servers in Oklahoma. Few of the tortillas yielded significant amounts of bacteria, but if anything, the ones from gloved workers were more likely to harbor germs. The problem, apparently, is that one almost always contaminates gloves when putting them on. Like Pigpen in the *Peanuts* comic strip, who was always surrounded by a cloud of dirt, we all go about our lives in a miasma of microorganisms, and we may as well get used to the idea.

CHAPTER 3
NOT SUCH A BAD CASE

The headlines are full of dire warnings about avian flu, and scientists are charting its progress in birds from the poultry markets of Asia to migratory waterfowl in Europe. But while the economic impact of a disease that kills birds is serious, that is not what keeps public health officials awake at night. They are not even primarily worried about the much-touted prospect of the flu mutating to a human disease. The real problem is how severe the human version will be, what scientists call its virulence. Some flus are only a little worse than a bad cold unless they strike infants, the elderly, or the immune-impaired, but others, like the one that caused the 1918 pandemic, can kill millions of people, even healthy adults. Which pathway will avian flu take?

To answer that question, we need an evolutionary approach. Why are some diseases mild, like colds, while others are devastating, like Ebola or SARS? Other diseases are brutal in one population but manageable in others, like measles, which slaughtered nineteenth-century Native Americans but is a comparatively mild childhood disease in people of European origin. If we can understand why virulent diseases get that way, we can better control them and predict when they are likely to strike.

———

A WELL-TEMPERED PARASITE:
THE EVOLUTION OF VIRULENCE

Parasites are organisms that live their lives on or inside another organism, called a host, and use the host's resources for their own benefit. To qualify as a parasite, you generally have to do your host some harm. But how much harm differs dramatically from parasite to parasite. A parasite's virulence refers to the amount of damage it does to the host, ranging from slight, in the case of a cold or the presence of a few head lice, to severe, in the case of Ebola virus or dysentery. Yes, a cold; by this definition bacteria and viruses can claim to be parasites with as much justification as the average worm, and all three, along with one-celled creatures like amebas and a few other oddities, can also be said to be pathogens, or disease-causing organisms, even though the nature of those diseases differs dramatically.

For many years, the conventional wisdom among biologists held that a proper parasite evolved to become relatively benign; why, they reasoned, would a parasite bite off the hand that feeds it, not to mention the vital organs making that feeding possible? The idea was that only those diseases that found their way to a new host would run rampant and exert unintended ill effects, as if they were overenthusiastic puppies who got into the yard and dug up the roses. Once the disease organism settled down and learned to behave, as it were, these regrettable excesses would be moderated and host and parasite would live together, if not in equanimity, at least without total ruin. A. L. Baron typified this view in 1958, musing, "If we have correctly interpreted nature's law, then all our disease germs will change from antagonism to coexistence, and turn from dangerous bits of alien life into inconspicuous particles within our living cells. Perhaps far ahead in the future in a disease-free world, the descendants of germs and men will live together harmoniously in a mingling of proto-plasm—a perfect symbiosis of men and germs."

It certainly seems logical that only a foolish parasite would demolish its host. While thinking that devastating diseases would become innocuous in the long term doesn't help current sufferers, it does provide some reassurance about the ultimate fate of humanity. It also suggests that if a disease is introduced to a new population, as when Europeans brought measles to the New World, it is likely to be deadly, but that in time, as the pathogen becomes accustomed to its new host, these lethal ef-fects are sure to diminish. Thus, the only good disease is an old disease, and public health efforts should focus on new introduc-tions, not on diseases that have been around for a long time.

The problem is that this conventional wisdom bore a strong resemblance to wishful thinking. It is similar to equally outdated mid-twentieth-century ideas about the balance of nature and so-called "prudent predators." Ecologists at the time, as well as Dis-ney movies about nature, portrayed predators like lions or wolves as wildlife managers, carefully cropping the herd of gazelles or deer and never overhunting their prey. As with the benevolent parasite, it seemed reasonable that such predators would reap the long-term benefits of never running out of food. From the prey's perspective, as from the host's, this seems like at least some consolation.

Both notions, however, are wrong, or at least inaccurate por-trayals of what happens in the course of evolution. To see why, remember that predation, like disease, is a two-way street, with both parties contributing to the outcome. In a population of wolves consisting of prudent individuals that heroically refrain from killing every prey animal they can, imagine that a newcomer arrives. This interloper has no such compunctions, and for its

own short-term gain it eats all of the prey it can catch, with no regard for the prey animal's state of frailty. That interloper will gain
more offspring, offspring that inherit their parent's rapacity. It is
easy to realize that such a strategy will spread more quickly
through the population than the more restrained one, resulting
in a group where it is every wolf for him or herself, at least from
the standpoint of exploiting the deer. Even if in the long run the
deer go extinct, the rapacious mechanism for sustaining their
populations will still be favored in their predators. Because it has
no foresight, nature is not particularly balanced, and what stands
in the way of doom for the prey is not restraint on the part of the
predator but selection for ever swifter deer.

Exactly the same what's-in-it-for-me principle applies to
disease. Again, imagine that a type of bacteria exploits a population, this time of people. Say that the parasite is spread when a
mosquito first bites an infected individual and then proceeds to
bite another, transferring bacteria in the process. Now compare a
bacterium that causes its host to become lethargic and sleepy, ne-
glecting the chores of daily life and lying around under a tree, to
a bacterium that has relatively little effect on its host. Which host
is more likely to be bitten by a mosquito and transmit the disease? The sicker one, clearly, since she is less likely to swat the
swarms of insects, have the derring-do to put screens over the
windows and nets over the beds, or act to deter attack. So increased virulence in this case favors the parasite and there is no
reason to expect the bacteria to become more benign. If, on the
other hand, the bacteria is spread not by a mosquito but by direct
contact between infected and uninfected individuals, a host who
feels only mildly ill is more likely to go out and spread germs
along with a handshake than one who is so debilitated he can
only lie in bed moaning gently and calling for ginger ale and

magazines. In this case, a less virulent pathogen does better. Either way, the old adage about things getting worse before they get better doesn't apply. Sometimes they get better, and sometimes they only get worse. It all depends on what makes the parasite live long and prosper, or at least live long enough to pass its offspring to a new host.

This evolutionary line of reasoning was laid out in the 1980s by Paul Ewald, an evolutionary biologist at the University of Louisville in Kentucky. Although he started his career watching the behavior of hummingbirds and sparrows, Ewald soon became convinced that that we could use Darwinian medicine to predict whether a disease was likely to become more or less virulent, and furthermore, to suggest that the conventional wisdom about the good parasite needed to go the way of so many other trends from the 1950s. He championed mode of transmission as key—as outlined above, diseases that are spread by intermediary animals like ticks or mosquitoes, called vectors, are likely to keep their victims debilitated, while those that require a helping hand to get from one host to another will tend to evolve a more innocuous way of life. He then went further, suggesting that once we understand how the mode of transmission influences virulence, we can take things into our own hands and actually direct diseases to become less harmful. His work both on this topic and on the infectious nature of many systemic diseases like cancer has been extremely controversial, but it sparked a debate about disease dynamics that is still raging. For example, will AIDS, even without the advent of more powerful drugs, become a milder disease in time, one that is manageable but not a death sentence?

Scientists are still arguing about the best way to test Ewald's idea. But it turns out that one had already been done, and in a place far from the hospitals and universities of either North America or Europe. It involves a disease not of humans but released by them, in an attempt to control yet another human intervention that had gone terribly wrong.

CHAINSAWS OF THE OUTBACK

Rabbits seem like innocuous animals, household pets for some, harbingers of spring and chocolate for others. To many Australians, however, they represent environmental devastation of the highest order, as testified by the subheading above, taken from a report by CSIRO (Commonwealth Scientific and Industrial Research Organisation, one of the main Australian government research groups). Rabbits are not native to Australia, but along with many other European animals and plants, they were deliberately introduced by people trying to re-create their English homeland. Although a few rabbits arrived with the First Fleet in 1788 and some pets escaped over the next seventy years, the real invasion occurred in 1859, when a farmer in Victoria named Thomas Austin sent for twenty-four wild rabbits. Austin wanted the animals for hunting, and his wish was realized beyond his wildest dreams. By 1865, 20,000 rabbits had been killed on his property, and of course the remainder flourished and bred, naturally, like rabbits, this being one case where a cliché actually holds true. They and their descendents sped across the continent at a rate said to be unparalleled in the history of such invasions, with some observers almost skeptical that the creatures could have managed to find enough time to stop and give birth to a litter of young.

Rabbits eat plants, and while some compensation for the damage they did to both natural and agricultural vegetation existed in the form of profits from meat and fur, it wasn't nearly enough, and they were deemed a major pest by the 1880s. They also contributed to soil erosion as well as the extinction of many animals and plants found only in Australia. Attempts to control their numbers, or at least their movements, included the construction of purportedly rabbit-proof fences that extended from one end of the country to the other; the aboriginal children in the 2002 film *Rabbit-Proof Fence* followed one of these for 1,500 miles. The rabbits themselves were undeterred, burrowing underneath the wire, and in the 1940s numbered in the hundreds of millions. A somewhat offbeat effort to make such a large number comprehensible came from the Sporting Shooters Association of Australia, which calculated that at an efficiency of five minutes and one shot to kill each animal, about $6 million of ammunition and 950 round-the-clock person-hours of hunting would be required to eliminate the creatures.

In 1888, a hefty reward was offered to anyone with a means for controlling the rabbit populations; entries were received from all over the world, including one from Louis Pasteur, but none were deemed acceptable. A viral disease called myxomatosis, a smallpox relative that infects rabbits in other parts of the world, was suggested as a control measure as early as 1918, but authorities were initially leery. Finally, though, the disease was released and took hold in 1950. Like malaria or yellow fever, myxomatosis is spread from one individual to another through a fly vector, which transmits the actual disease-causing organism—the virus in this case—in its bloodstream. When the infected insect vector bites its victim, the pathogen is spread. Heavy winter rainfall at the end of that year helped the carrier mosquitoes breed, and within three years, rabbits in all parts of Australia were infected.

But a few years later, a funny thing happened. Instead of killing all the rabbits that became infected, after a few years the virus started killing only 95 percent of them, a decline that doesn't sound like much but which has significant consequences. In addition, sick rabbits took three or four weeks to die, instead of the week to ten days they had previously. The effectiveness of the pathogen continued to drop, and today, only about half of an infected rabbit population will die of the disease. Australian scientists have turned to other control measures, including other viral diseases, because myxoma simply isn't enough.

Why did the rabbits survive? From this particular virus's perspective, not killing its host immediately is beneficial, because a more mildly affected rabbit is more active, allowing more insect bites and further spreading the disease. Viruses vary in their genetic composition just like any other organism, but instead of traits such as height or intelligence, viruses differ in the severity of their effects. In this case, less severe strains or subpopulations of the virus are more likely to spread, and hence the disease becomes less virulent. From the perspective of the rabbit, the relevant trait is a natural resistance to the myxoma; again, individuals vary, and by chance some rabbits were able to defend themselves, at least temporarily, against the virus. Such rabbits had an advantage over those that succumbed more readily, and they too enjoyed a reproductive upper hand. The combination of reduced virulence and enhanced resistance meant that the myxoma lost its edge.

The myxoma story is often held up as proof of the "good parasite" idea, because indeed a new pathogen evolved to become more benign in time. The key point, though, is that the virus did not become completely harmless. Instead of relentlessly marching to ever-lower virulence, it stalled at a moderate level. At the current degree of virulence, myxomatosis is not a hop in the park for rabbits, and still kills a substantial proportion of those infected. And experimental work with mild lab strains of myxoma suggests that the virus *could* have evolved much lower levels of virulence. In fact, once host resistance had evolved, the virus actually evolved back toward higher virulence. Establishing such a "happy medium" level of virulence makes sense when you look at this trait from the parasite's perspective.

AS BAD AS IT GETS

But can we use information about what favors virulence to control diseases and render them less deadly? Ewald considered two important aspects of transmission: first, whether a disease is vector-transmitted, like myxomatosis, or is transmitted directly, through contact with a germ-laden hand, for example. Vector-borne diseases can often afford to be virulent, because debilitating their hosts just makes those hosts all the more vulnerable to attack by insects. Diseases that are carried in the water supply, like cholera, are similar to vector-borne illnesses in this respect, despite being classified as directly transmitted, because even when the victim is incapacitated, the pathogen can be spread when, say, infected clothing is washed in a communal water source.

Second, Ewald pointed out that diseases that are spread primarily from mother to offspring, a process known as vertical transmission, are likely to be less severe than their counterpart, horizontally transmitted pathogens. Some human pathogens, such as the hepatitis C virus and HIV, can be transmitted from a mother to her baby, but the most common vertically transmitted diseases occur in nonhumans, particularly insects. The West Nile virus pathogen, for example, can be passed from a female mosquito to her eggs, an important consideration when mapping the potential spread of the disease. A host that is debilitated to the point of being unable to mate and reproduce will also be unable to pass on the disease. A survey of worm parasites of wasps showed that indeed, the species of parasites that were less deadly were also more likely to be transmitted from mother to offspring, and that furthermore, the species of worms that were the most harmful to the wasp had just as old a history with their host as the milder types, contradicting the idea that novel diseases are the most severe.

Ultimately, a parasite must balance costs, in the form of killing off a host it can still exploit, against benefits, in the form of reproducing and getting to a new host more readily. A pathogen that is highly virulent is exploiting its environment in a reckless, party-like-there's-no-tomorrow way; it uses up one host and travels quickly to another, and another. Having a vector means you can ditch a host without a backward glance.

Still another way to remain virulent is to be able to last for a long time in the environment without being inside of a host, like anthrax, which survives for years in a spore or seed-like form in soil, or smallpox, which can last more than ten years outside of any human. Those pathogens can afford to be damaging because if they do kill their hosts and find themselves homeless for a period, they are able to weather the hiatus and wait for a new victim. In contrast, the vectors of a disease themselves should be unlikely to get very sick from the ailment they carry, since although it's hard to imagine a feverish mosquito, anything keeping the messenger of illness from her appointed rounds would also harm the pathogens themselves. In a similar manner, diseases that infect prey animals can evolve to be more virulent than those infecting predators, particularly when diseases can infect multiple species, since a sick predator will have difficulty getting food, but sick prey are easily eaten.

From this common-sense starting point, Ewald and others have championed the idea of virulence management; if we know that certain things favor reduced virulence, let's manage the environment of an illness so as to sway it in that direction and thereby soften its effects. For example, cleaning up water sources can prevent such passive transmission via contaminated dishes or bedding, forcing the disease organism to be transmitted from people who feel well enough to do something other than lie in bed or run to the toilet. This means that the disease must become less virulent and have less impact on its hosts if it is to continue reproducing. Other forms of management include blocking vector access to moribund hosts by putting up mosquito netting and keeping insects from biting them as easily. This too encourages the spread of less damaging forms of the disease.

This idea has a great deal of intuitive appeal, and is based on sounder evolutionary theory than the notion of either prudent predator or good parasite. Few if any evolutionary biologists continue to support the old premise that virulence is just a by-product of a new disease and that all diseases will eventually equilibrate to a more benign state. Dieter Ebert of the University of Fribourg in Switzerland has done experiments with parasites of *Daphnia*, tiny crustaceans sometimes called water fleas. These as well as his surveys of the scientific literature concur that on average, parasites are most virulent in the host where they originally evolved, not in the hosts in which they are recently introduced. True, notable exceptions exist, like Dutch elm disease, which ravaged the elm forests of North America, or rinderpest, a disease of cattle and other grazing animals that has likewise decimated herds in Africa, but these get a lot of attention precisely because they are exceptions, kind of a man-bites-dog phenomenon. The hundreds of cases in which diseases are introduced and fail to have any effect on the local host population are simply not newsworthy.

But Ebert, along with University of Texas biologist Jim Bull and several other scientists, is cautious about making recommendations for disease management in the real world. They watch with alarm as Ewald makes recommendations for health policy, given the limitations of extrapolating from experiments on water

the local concrete salesperson than a high school teacher. Unless of course, it is parent-teacher night at the school, in which case . . . well, you see the point. Diseases are not bouncing around at random from individual to individual, but instead are selectively borne from place to place depending on the encounter rate of those in the population. Therefore, you can't assume that a disease will easily travel to all members in a group, and calculating disease transmission based on equivalent movement within the entire population of New York City, say, will give you faulty results.

What's more, the same pathogen can have different virulence at different times. A gut parasite of bumblebees called *Crithidia bombi* is generally quite mild in its effects, and up to 80 percent of the bumblebees in Europe will have an infection at any one time. However, Mark Brown, working in the laboratory of Paul Schmid-Hempel in Zurich, examined the bees at different stages of their lives. Bumblebees hibernate during the winter, and during that time the bees, like miniature bears, depend on stored fat reserves to carry them through. If a bee has had a lean season, such reserves may not be enough, and so hibernation is a physiologically stressful time. Like the more familiar honeybee, a bumblebee colony has a queen, but the queen starts her colony anew each spring, and she must, like any working mother, juggle numerous tasks, simultaneously producing wax, rearing the new brood, and keeping her ovaries in high gear. Similarly, then, colony-founding is an energetically draining business. Brown and Schmid-Hempel, working with Paul's wife, Regula Schmid-Hempel, found that the parasite had much more severe effects on bees at these demanding times of life than during more tranquil periods. An infection that's trivial during the idyllic summer days of gathering pollen can mean the difference between life and death when snow is thick on the ground. Such context-dependent effects, as the researchers called them, are probably more common than we realize, and they further complicate our ability to domesticate disease.

THE NEW FLU: PLAYING CHICKEN?

Let us go back to avian flu, perhaps the most recent in our worries about a new hypervirulent disease sweeping through the planet. What do evolution, and flu's mode of transmission, teach us about the likelihood of such a human pandemic? First, the basics: Flu comes in different strains, and our human influenzas often come from other animals, notably pigs (including the "swine flu" of 1976), but also birds, like the strain of current interest, H5N1. Sick birds are one problem, but the real concern is that H5N1 will do two things. First, that it will mutate so that it is passed from person to person, not just from birds to people, as it does now. Second, that if human-to-human transmission does happen, the mutated form of H5N1 will be extremely virulent, like the 1918 flu virus, which killed anywhere from 20 million to 40 million people. The 1918 virus and H5N1 are genetically similar, which fuels this latter fear.

How realistic is this scenario? Wendy Orent, author of *Plague: The Mysterious Past and Terrifying Future of the World's Most Dangerous Disease,* points out that the modern avian flu is in large numbers of poultry, not wild birds. If wild birds got virulent forms of flu, they would be so sick they couldn't fly, and if they couldn't fly, they couldn't effectively transmit the disease; it isn't in the best interests of the virus to be deadly. Some wild birds with flu have been found, but they do not suffer from the disease in anywhere near the huge proportions that poultry do. Avian flu in Asia is promulgated in large part because of the astronomically high levels of crowding in commercial poultry operations, which allows extremely rapid spread without the usual constraints of having to travel from one mobile animal to another. A chicken farm is like a buffet for viruses. So the likelihood of mass transmission via migratory wild birds is slim, which is a good thing, in my opinion, given the recommendation by a call-in listener to Orent who suggested poisoning the food used by migrants as they journey between continents. Orent herself, whose book title shows she is clearly not a slacker when it comes to foretelling apocalyptic disease scenarios, suggests caution in predicting another pandemic. It is possible that wild birds could carry the disease without being affected, and then spread it to humans, but this is unfounded to date.

Well, but what if the flu moves from poultry, rather than wild birds, to humans? Here too, mode of transmission matters. Ewald believes that the flu was unusually virulent in 1918 because of extreme crowding in the trenches and transports of World War I. As Orent puts it, "The Western Front was a disease factory, and it manufactured the 1918 flu." Even the most crowded slum does not replicate the conditions of the soldiers. Under the cushy (for the virus) conditions of cheek-by-jowl contact among men at war, there was no penalty for being lethal, the way there would be for a flu virus committing immediate host-icide under more normal circumstances. After the war ended, the flu became far more benign, as its luxurious lifestyle was eliminated and it had to go out and hustle for new victims who could, though feeling ill, still stagger to their workplace and sneeze on unsuspecting coworkers.

Further reassurance comes from two independent 2006 studies that showed that H5N1, unlike flu viruses that are already in humans, lodges not in the upper respiratory tract but in its lower reaches, where the cells deep in our lungs are similar to respiratory cells in birds. This makes the virus difficult to transmit via the usual cough or sneeze.

No one can guarantee that the current avian flu absolutely

won't mutate to a deadly form in humans. And virtually all researchers agree that it is prudent to take steps to ensure that we would be ready to face an epidemic should one arise. But to use Orent's words again, "the flu virus, like all of life, is subject to evolution. Lethal diseases don't fall out of the sky."

CAN'T TOUCH ME

It might seem that the best way to keep from worrying about how virulent a disease is or might become is to get immune to it altogether, either by getting ill and recovering, hence developing a natural immunity, or through vaccination. But it turns out that immunizing people actually affects the way the disease itself evolves.

How does vaccination work? Any foreign invader in the body, whether it is a bacterial cell, a worm egg, or an injection of killed smallpox virus, alerts the immune system that something nonself has arrived. An elaborate sequence of events then results in the production of antibodies, molecules that recognize the foreign objects and help other parts of the immune system destroy them. Once a host encounters a disease, and has manufactured antibodies, subsequent encounters with the same pathogen will be met with an already-assembled defense, and the disease will be stopped in its tracks. Vaccination mimics the triggers of the immune response to a particular pathogen, but does not cause the disease. Not all diseases can be stopped this way, but for those that can, the immune system—and vaccination—is literally a lifesaver.

But while the body's immune defenses are at work, the pathogen is continuing to reproduce, and that means it can still evolve. So a disease's virulence can change in response to the immune defense mounted to respond to it. For instance, the immune response often takes several days to develop, which should encourage rapidly developing, and hence more virulent, pathogens. Or a pathogen could evade immune recognition by changing its appearance during the brief window while antibodies are being produced, and hence maintain a higher level of virulence than it could otherwise. Evidence of the latter is seen in the parasite that causes African sleeping sickness: Its protein coat can undergo many thousands of alterations in a few days, effectively eluding the immune system.

These ideas seem reasonable in theory, but does virulence really change simply as a result of immune response? To put the theory to the test, Margaret Mackinnon and Andrew Read from the University of Edinburgh used laboratory mice and malaria in a kind of reverse vampire experiment, in which they repeatedly removed blood containing the malaria parasite from the mouse hosts and injected it into two other groups of mice. One group of hosts had been immunized against malaria previously, and the other, called naïve, had not. Mackinnon and Read wanted to see if the host immunity influenced the evolution of the disease. As it turned out, it did: When the evolved lines of parasites were injected into new groups of immunized and naïve mice, the disease was more virulent if it had been passed through immunized mice than through naïve ones. Furthermore, when the researchers allowed malaria to be transmitted from mouse to mouse via the usual mosquito route, overall virulence dropped, but the immune-selected lines of parasite were still more virulent to the naïve mice.

Mackinnon and Read's experiment sounds esoteric, but in fact they were imitating a common practice in vaccination by mimicking what is called an "imperfect vaccine." In other words, the population of mice hadn't all been given the same protection against the disease. Vaccines are imperfect if they either don't confer complete protection, so that some of those vaccinated still get sick, albeit not as severely, or if their protection fades after a period of time, so that individuals go from being immune to being vulnerable again, as may be the case for smallpox. Mathematical models describing the fate of different types of vaccines in hypothetical populations suggest that vaccines designed to slow down the growth of pathogens within an individual host or reduce their toxicity to that host also allow the more virulent strains to persist, so that unvaccinated individuals get hit harder by the disease. This seems to be what happened to Mackinnon and Read's mice. On the other hand, vaccines that actually reduce the likelihood of getting infected in the first place cause less virulent strains to be successful, and may also mean that the disease can be eradicated more easily.

These considerations are being taken into account as medical researchers and public health workers struggle to develop vaccines to HIV and malaria in humans, both diseases for which vaccines are likely to be imperfect. They also serve as a cautionary note for those parents itching to take their children off the vaccine treadmill. Vaccination is effective in reducing or eliminating a disease in society because once nearly everyone is immunized, not enough susceptible individuals remain to transmit an infection, even if a few people with the disease enter the population. Of course, the risks are not zero, because vaccines do have side effects in a small number of people, and in a fraction of those, the side effects may be severe. They are far outweighed by the consequences of actually getting the disease rather than the vaccine, but they still exist. Therefore from any parent's or child's perspective, the best thing is to be unvaccinated in a vast population of compliant individuals, so that one benefits from everyone else's immunity but does not pay the price of being inoculated oneself. Needless to say, such a dream is not likely to be realized, since all the other parents would want the same thing.

The increasing numbers of parents opting out of vaccinations thus present a huge risk for everyone else, because if more than a tiny fraction of individuals is vulnerable to infection, the disease can spread with enormous ferocity, and it is much harder to control an outbreak than prevent one. Modern parents lack memory of the devastating effects of a common disease like measles, which can cause encephalitis, leaving a child blind, deaf, or retarded, and it can also cause death, particularly in developing countries, where malnutrition and vitamin A deficiency are prevalent and as many as one out of four people die from the disease. Even in industrialized countries, one or two out of every thousand children infected with measles dies.

Despite all of these caveats about the complications of predicting virulence, few people would argue against Ewald's proposals for virulence management, which feature cleaner water and better sewage disposal. And some scientists say that Bull, Ebert, and other critics are simply expecting too much from a modest framework that was really only intended as a first step. Even if we can't implement virulence management for every disease, or even for very many of them, this is often because we lack basic information about the lives of the pathogens, not because the theory itself is faulty. Evolutionary biology is still our best shot at figuring how bad a disease is likely to get.

CHAPTER 4

THE RACE WITH SEX THAT'S NEVER WON

"Love is the answer, but while you are waiting for the answer, sex raises some pretty good questions."
—*Actress Glenda Jackson*

Never mind the question of why a woman can't be more like a man. What we really need to know is why men and women both can't be more like geckos. Geckos, those small tropical lizards with the pads on their toes that enable them to run up walls in search of flies to eat, are endearing enough to be featured on Hawaiian T-shirts and other vacation paraphernalia. But it isn't their cuteness we should aspire to emulate; it is their mode of reproduction. In many species of geckos, males are nonexistent. Females lay eggs that hatch into other females, each one genetically identical to her mother before her and her daughters to come. If you look at the belly of one of these geckos as she clings to the window screen of your Hawaiian hotel room, you can even see the eggs gleaming softly through her translucent skin. It is as if she is about to produce her identical twin. She has never had to worry about finding a mate, about whether her sons will successfully compete with other males. There is no need for them to compete because there are no males; like a science fiction character, she produces one perfect daughter after another. She and her descendents form a clone, produced without fanfare or governmental regulation. Just as it is possible to have sex without reproducing, it is possible to reproduce without sex.

Though it is scattered throughout the animal kingdom, this absence of sex is rare, with such an admirably chaste existence practiced by fewer than one in a thousand of the world's species. Everyone else has to deal with reproducing sexually, with all the accompanying complications and consequences of dating: heartbreak, the risk of sexually transmitted diseases, wedding registries, in-laws, and deciding who gets up when Junior cries at night. If other species don't have these exact same problems, they have versions of them, in the form of growing that big tail that is de rigueur for a sexy male peacock, or finding a mate that will bring dead animals to the lair for you to eat if you are a fox.

If the idea of being a gecko is unappealing, other possibilities include tiny crustaceans called water fleas, several kinds of insects ranging from moths to grasshoppers, desert lizards called whiptails, and a smattering of others across the animal kingdom. Still other species, like aphids, give birth to daughters without sex some of the time, and then mate and produce sons the rest of the time. Look closely at the green ovals on your roses and you will see at least a few of them with even tinier green blobs protruding from their abdomens. These female aphids have never seen a male, but they have babies just the same.

The question of why we all aren't like geckos or aphids goes even further. Sex isn't only cumbersome; it is also senseless. At first glance, from an evolutionary standpoint, sex should not exist at all. Asexual reproduction, the manufacture of copies of oneself without recourse to any other individuals, is far more efficient. Go back to that science fiction, and say that aliens from outer space invaded Earth. Imagine that they are like us in every respect, except that the extraterrestrials are all women that can reproduce asexually. In a city the size of Detroit, the aliens would replace us sexual humans in less than fifty generations, even if they started with only a single female. This assumes that the population is at a stable size, so every human couple has exactly two surviving children and the invader has just two daughters, each of which has two of her own.

To picture this more clearly, compare the alien Zara to human Mary Sue. Mary Sue has a son and a daughter, Herbert and Melanie. Zara has two daughters, Zorann and Zelia. Zorann and Zelia then each have two daughters of their own, but Herbert has to find a mate and can't produce any offspring except by finding a woman to bear his children. Melanie has her requisite two babies, but Mary Sue is already outnumbered by Zara in the number of grandchildren she has. Keep this up and the end result is the elimination of humanity as we know it, just like in the movies, in far less time than it took us to get from the Bronze Age to Google. Biologists call this the cost of producing males, referring to the problem of sexually reproducing females producing sons, who don't have any progeny. It's not that the males don't contribute anything; they do, in the form of genes. But those genes have to join with a female's genes in order to do any good. It's as if instead of buying one and getting one free, you have to produce two to get one in return.

Another way to look at it highlights the reason evolutionary biologists are so fascinated with why we bother with sex. Zara's daughters are all exactly like her, so that Zara's genes have a 100 percent chance of being passed on in succeeding generations. But Melanie and Herbert are only half like Mary Sue, because Mary Sue needed the sperm from their father to fertilize her eggs. Any egg cell, like any sperm cell, only has half the genes of its parent. A gene thus only has a 50 percent chance of being passed on in an offspring that is produced via sex. This degree of relatedness trickles away every generation, so that a grandson is only one quarter like his grandfather, a great-granddaughter only an eighth like her great-grandparents, and so forth. Success in evolution requires that an individual's genes get passed on to future generations, and asexual reproduction does this far better than the messy and wasteful business that is sex.

All of this means that Zara and her compatriots ought to find Earth pretty easy pickings. Luckily, they or beings like them don't seem to have discovered us. But do not rest easy just yet. The equivalent of the asexual being from outer space is already here, in the form of all those animals like the geckos. Oddly, though, the geckos have close relatives that do reproduce sexually. Why haven't the asexual families swamped out the sexual ones? In a kind of snail that lives in pristine lakes in New Zealand, sexual and asexual populations live virtually side-by-side. Snails who have sex shouldn't stand a chance next to their all-female competitors. And of course, the snails are only one case of asexuality; if it can occur in so many different animal groups, why hasn't it caught on like wildfire? In short, why is sex so common?

This question has been called "the queen of problems in evolutionary biology" by Graham Bell, a British biologist who has spent much of his life pondering it. And it is important to recognize the realistic implications of why sex exists among most organisms; it's not one of those angels-on-a-pin problems that academics are often accused of wasting time on. If we can understand how sex evolved, we can better understand how genetic information is processed and how genetic "mistakes" are naturally corrected, both essential to diagnosing and curing genetic disorders. But most intriguing, the story of how sex evolved is the story of why every individual is unique.

The answer to all of these questions, it turns out, lies at least partly, and maybe mostly, with disease. We make much of our uniqueness—it is at the core of our being. If parasites explain why no two of us have the same genes, then they may also partially explain why a drug that works for our aunt may not work for us, why two children learn math differently, and why some people like poetry and others poker. Parasites have made us not only male and female, but also into creatures that differ one from the other in ways that make the proverbial snowflakes look like photocopies.

SPLITTING HAIRS AND CELLS

To understand why parasites are so critical to the evolution of sex, we first need to ask, Clintonesque, just what sex is. This is not as facile or pedantic as it might appear. The word "sex" can mean three different things: reproduction, genetic recombination, and gender. The first simply means the manufacture of more individuals, something that, as we have seen, can occur with or without a male and female joining their genetic material to produce offspring. Gender refers to the differences between males and females that go beyond simply an X chromosome for girls and Y for boys, like the bright colors of cardinals or the enormous tusks of elephants.

Genetic recombination, the second part of sex that needs to be explained, is at the crux of why parasites caused us to have sex. Genetic recombination is the mixing of genes that happens when sperm meets egg. Sperm and egg cells have only half of the genetic material of their parents, which is why offspring are only half like either their father or mother. But if a straightforward splitting of parental genes was all that happened, then every sperm cell would be like every other, as would every egg, and every child from a given set of parents would be exactly like every other child, although they would all still be different from their parents. As siblings know, however, this isn't the case. Each child inherits a different palette of genes from each parent, so two levels of genetic mixing occur with each act of reproduction. Reproduction is more than just using two halves to make a whole; it is making many different kinds of halves to begin with.

How does this rearrangement happen? Humans have forty-six chromosomes, and each one occurs in a duo with another, so that we have twenty-three pairs. To make a sperm or egg cell, the paired chromosomes in the parent cell have to line up and split. While they are paired, but before they part company, a curious process takes place. Genetic material from one chromosome is swapped for the complementary genetic material on another chromosome, thus yielding new combinations. Imagine a folk dance in which all the girls are lined up opposite all the boys. The caller for the dance says, "Switch places!" but only some of the children pay attention and move to the other line. When the lines separate, they are each mixed-sex.

To apply this to genes, say that a father has genes for brown hair and blue eyes on one chromosome, and genes for blond hair and brown eyes on the other chromosome. The father himself has brown hair and brown eyes because those traits are dominant. In other words, their effects are shown while the effects of the other genes are not, a fairly common situation for many traits. Without recombination, all the man's children would either inherit the brown hair–blue eyes chromosome or the blond hair–brown eyes one. But while the chromosomes are lined up, they form temporary connections and swap genes, so that the gene for blond hair and the one for blue eyes can end up together on the same chromosome. If the sperm cell bearing that chromosome fertilizes an egg, the resulting child could end up with blond hair and blue eyes, depending on the mother's genes. A different sperm cell has the brown hair and brown eye genes together, and will produce yet a different child.

This is all well and good, and certainly gives fodder to doting parents who are convinced that their little darlings are wonders never before seen on the face of the earth. But why should shuffling genes around be part of sexual reproduction? And furthermore, how does the generation of variety help in the competition against Zara, busily cloning herself in Detroit? Herbert and Melanie are different from each other and from Mary Sue, bless them, but there is still that cumbersome need to combine two sets of genes to make one. How can new sets of gene combinations help fight the relentless wave of identical asexual beings?

A GENETIC NEED FOR SPEED?

Recombination has been understood for over a century, and biologists have been wondering about the utility of sex for at least that long. One possibility is that sex evolved to clear out the mutations that accumulate when DNA replicates asexually. Because the genes realign on their chromosomes every time a male makes sperm or a female makes eggs, the number of mutations that have built up diminishes, depending as they do on the association of a particular set of genetic components.

One of the earliest attempts to explain the evolution of both sexual reproduction and recombination was that those phenomena gave natural selection a constantly changing pool of individuals. Therefore, the reasoning went, evolution could proceed more rapidly and species could change more readily as well. Sex, then, helped us get from sea to land, and ape to CEO, without wasting too many millennia in the process.

This idea is intuitively appealing, particularly to us Westerners who value efficiency and speed. But the need for fast evolution isn't a good explanation for sex, for two reasons. The first is that faster evolution, regardless of how laudable it may seem, is not a necessity for nature. Changing rapidly as the epochs go by doesn't make a species more likely to succeed; plenty of organisms look just like they did eons ago and get along just fine, thank you. Cockroaches, for example, appear in hundred-million-year-old fossils looking eerily as if they could have been scraped from someone's kitchen floor.

The other reason that this explanation is wrong is that it relies on a benefit, not to an individual, but to the species as a whole, to allow it to persist. As with the prudent predator, these good-of-the-group arguments can't work for sex either. Consider the supposedly suicidal penguin. Some kinds of penguins live in places where voracious leopard seals, so named because of their spotted pelts, roam the waters looking for prey. A penguin on land is safe, but of course penguins have to get into the water to find their own food. Penguins often enter the sea in groups, scanning for seals in the water below. If they spot one, they enter at another location. The problem is that the best way to determine if a leopard seal is lurking is to send in a voluntary sacrifice, a test penguin that everyone else can watch. If the first bird in the water surfaces unmolested, everyone else can jump in, but if the sea suddenly roils with a bloodstained struggle, well, maybe everyone should move to a cliff further down the coast.

But this system can't be maintained via natural selection. Imagine two kinds of penguins, the heroic leapers into the water and the prudent hangers-back. The leapers leap, the hangers-back hang, and soon the population is filled with penguins waiting for the next guy to get into the water. They will wait a long time, however, because the genes for those sacrificial tendencies got eaten, along with the heroic tuxedoed bodies they resided in, by the leopard seals. Preserving one's own genes wins, even if in the long term that means that the genes of the entire species disappear.

What, then, are the penguins doing when they appear to send out the noble sentry to draw the seal's fire? A good guess is that they are shuffling at the cliff's edge until an unlucky loser falls in, and since falling in didn't depend on the faller's genes, but on random chance, the system is maintained and selfish behavior still rules. Virtually every other story about things happening for the good of the group or the survival of the species likewise self-destructs if you imagine similar contrasts between the selfish and altruistic members of the society.

For the same reason, sex cannot have evolved like a lottery, a common analogy. Which would you rather have, twenty tickets with the same number or ten tickets with different numbers? Obviously the latter gives you a much better chance of winning. So producing variable offspring could be beneficial even if those offspring are fewer in number.

This sounds fine, but, as with the suicidal penguins, it falls apart under closer examination. The disadvantages of sex are simply too huge to be offset by a chance at winning in a random draw if what produces the winning ticket, what makes a unique offspring survive, is left to chance vagaries of the environment. Say Zara and her daughters, and their daughters, are all well suited to the Michigan climate, but global warming turns Detroit into the equivalent of the Sahara. One of Mary Sue's descendents, it turns out, tolerates these desert conditions remarkably well, and while the extraterrestrials are suffering from deadly skin cancer and trying to find iced drinks, the sexually produced child runs happily through the scorching sands of Woodward Avenue, thriving on little water and living long enough to potentially pass on some of Mary Sue's genes, however diluted by the passage of generations. But how does this lucky youth find a mate? We need many lottery winners, not just one, and furthermore they have to be of the appropriate age and sex to repopulate the city. Sure, it could happen. But there is no evidence it does.

THE QUEEN OF SEX

But the most crucial reason for sex is parasites and pathogens. Remember Zara and her endless parade of daughters and granddaughters compared to Mary Sue's variety pack of descendents. Even if we assume that after the climate changed Detroit to a desert, Mary Sue's relation found a mate and prospered, it isn't likely that the climate would seesaw from cold to hot and back again, much less that the kind of cold or hot would be different and require different adaptations for Mary Sue's genes to master. Thus even if sex gave Mary Sue an edge over Zara once, it's unlikely to give her the same advantage again. The physical environment doesn't evolve back at you.

The biological environment, however, can do just that. Never mind the weary or wicked—the ones who really get no rest are the healthy. Because natural selection acts on both the parasite and its host, any adaptation that makes an animal better at resisting disease will be met with a counteradaptation in the pathogen that circumvents the defense. Many evolutionary biologists speak of arms races between hosts and pathogens, and again the analogy is instructive here despite my general dislike of it. Arms races can, of course, occur without actual war, and can go on for years, perhaps indefinitely, while each side escalates its weaponry. Similarly, hosts and parasites can and do continue to evolve responses to each other without either partner achieving a final victory. Host-parasite interactions provide the constant pressure to reinvent oneself, the perfect inspiration for sex.

This cyclical selection has been described in a hypothesis named the Red Queen, after the character in Lewis Carroll's *Through the Looking Glass* who pointed out, "Now, here, you see, it takes all the running you can do to keep in the same place." The "same place" is a certain level of resistance to disease, and the running is the continual cycling of new genes best accomplished via sexual reproduction. Imagine starting with a mouse host and its parasite, say a kind of worm that attacks the lungs of the mouse after being inhaled as a tiny egg. If a few of the mice happen to have lungs with a more rigid exterior that keeps the worm from burrowing in, the genes for such an ability will give their bearer an advantage, and they will become more numerous in the population. But inevitably, some of the worms will happen to have a way to thwart this, perhaps with mouthparts that can penetrate the armor. The worms then become more numerous, and the defense is useless. Then selection will favor not lungs with a protective coat, but noses with filters that keep out the worm eggs. A new gene becomes predominant. It too, of course, will be overcome by the worm, to be followed by yet another change in the kinds of mice in the population.

Where does this inexhaustible source of new gene combinations for both attacking and defending come from? Sex. Why does the host need to keep supplying them? Because the parasite, too, continues to muster its own new set of tools.

My students occasionally draw the hopeful but erroneous conclusion from this that an individual sex act will keep them from becoming diseased. Alas, this is not the case; we are talking here about events on an evolutionary time scale that happen in an entire population, not on a Saturday night date. But forms of the Red Queen hypothesis are increasingly accepted as an important part, maybe the most important part, of the explanation for the evolution of sexual reproduction in the vast majority of the plants and animals on earth.

TESTING THE QUEEN

If parasites have caused sex to evolve, we should be able to make some predictions about where those rare cases of asexual reproduction persist. Curt Lively, a biologist at Indiana University in Bloomington, has done some of the most compelling tests of the Red Queen hypothesis, using a species of conical snail only slightly larger than a grain of cooked rice that lives in beautiful mountain lakes and streams in New Zealand. Although the spectacular Middle Earth scenery is a side benefit, the real reason for his traveling to the Southern Hemisphere is that some populations of the snails reproduce asexually and are only female, while others have males and females, and these two types sometimes live in the same place. The snails also get infected by several different kinds of trematode parasites, worms that are passed from ducks into the water, where the snails acquire them. Lively therefore wanted to see whether the presence of the parasites had anything to do with the prevalence of sexual or asexual snails.

First, he and his colleagues examined the places where the parasites are most common, in lakes rather than streams. The sexual populations were much more likely to occur in those habitats. Even in separate populations within the lakes, the higher the proportion of infected snails, the higher the proportion of males. The idea that maleness accompanies worm infestations may be unappealing, but it is excellent support for the hypothesis that parasite pressure selects for sexual reproduction. Second, the different genetic types constituting the clones of asexual snails fluctuate over time, with some types being more successful when lots of parasites are present and others when fewer parasites occur, which is what you would expect if the kind of cyclical selection described above is happening. In fact, the genetic makeup of the snail seems to be important in determining whether it is likely to be infected, with populations of snails being best able to resist their "home" type of parasite but more vulnerable to parasites from the next lake over. Conversely, crossing two populations of the parasites from different places yields a hybrid that is not very good at infecting either of the snail types its parents could easily infect. The hybrids are fine, however, at infecting remote populations that have never evolved any resistance to them. All of this suggests that the snails and their worms have evolved an intimate, if inimical, partnership that depends on sexual reproduction.

Lively also does experiments in which he swaps the parasites from one location to snails from another, using aquaria to avoid destroying the ecological balance that exists in the natural bodies of water. He has found that the clones that were recently most common are now the most susceptible to parasites, which means that the parasite's genes are chasing the commonest host types, as the Red Queen hypothesis predicts. Sexual populations are better at resisting being overtaken by the parasite, but clones can persist at least for a little while, if they happen to have a resistant genotype.

A final odd twist on the story is that shallow-water snails seem to be more infected than those out in the middle of the lake. There are also more sexual snails in the shallows. It may be that the ducks that are the final hosts for the trematodes tend to feed and wade in the shallow water at the edge, providing a source of parasites that then go into the snails that live there. This sets the stage for the joint evolution of parasite and host genes that the Red Queen requires.

Sexual reproduction is not perfect, however. What about the completely asexual animals that seem to persist regardless of the virtues of the Red Queen, that tiny determined fraction of organisms blissfully reproducing sans males? Rotifers, microscopic animals that inhabit pond water, have, in the words of biologist Wayne Getz, "abstained from sex for 30 to 40 million years." This makes those teenagers vowing to remain chaste until marriage look like wimps. Mites, lizards, worms, fish—anywhere you look there are asexual species. A tenth of a percent of all the species that exist is still quite a few organisms, well over 100,000 no matter whose estimate of biodiversity you use. As I mentioned above, these are often living in so-called marginal habitats: windswept mountaintops, barren deserts, or Arctic meadows. Certain areas seem virtual hotbeds of celibacy; biologist Michael Kearney titled a paper "Why is sex so unpopular in the Australian desert?" Perhaps such habitats simply contain too few parasites to make the cost of producing males worthwhile, or perhaps the benefits of being able to multiply rapidly at the edges of sustainable life are worth the gamble of being infested. Or maybe those habitats do not change quickly, whether in the diseases they harbor or other components, and hence the pressure for sex and recombination is reduced. All of these ideas have found some support from researchers, though not all the answers are in.

Parasites are not the whole answer to why sex exists. Even Curt Lively, whose work probably goes further toward supporting their importance than anyone else's, believes that several factors, including the benefits of eliminating mutations, are likely responsible for the sexual reproduction being so widespread. But pressure from disease is the only solution that also explains at least something about why the asexual species exist where they do, as well as why sex itself could be advantageous.

———

DON'T BOTHER, THEY'RE HERE

This view of sex means that parasites, wholly or in part, are responsible for our existence as male and female, with all of the accompanying romance and politics. Without disease, there would be no sex. Not just a long dry spell, a contemplative time to explore the nuances of celibacy, but no *sex,* no defining ourselves in terms of gender and its stereotypes. Attempts in science fiction aside, I suspect we cannot imagine life without it.

Without disease, we would also not need to differ from each other. We could exist, like the geckos, as a series of exact duplicates, each like the other but unfathomably distant from the other sets of copies around us. Such an existence strikes me as rather lonely, lacking in connectedness, a connectedness that arises paradoxically from our differences. Doppelgangers seem intriguing because of their impossibility, but if there were no parasites, and no sex, they would be commonplace. You would not meet your double unexpectedly on the street in a faraway country, as it is portrayed in films; many of them would surround you as a matter of course, as you were produced and reproduced in turn.

One of my students used to wear a T-shirt that read, "It's not that it takes all kinds, it's that there are all kinds." This is a rather profound statement, really, and one that reflects the fodder for evolution that sexual reproduction and recombination yields. At any one time, we have no idea about which genetic combinations will be useful in the future, and indeed it is unlikely that all of them will be useful at once. But if the climate grows hotter, a gene for resisting desiccation will start to seem awfully handy.

Nevertheless, 99.9 percent of the organisms on the planet can't be wrong. Or at least they can't be altered lightly. After all, they evolved sex a very long time ago, anywhere from 2,500 to 570 million years ago, depending on whom you ask. And therefore if we try to be more like the geckos and carp, we are tampering with a process that has very old roots. The need to compete in that endless arms race is an ancient one indeed. This means that cloning sheep, not to mention people, can have unforeseen consequences. Most obviously, creating a society of identical people would make us more vulnerable to diseases. A similar problem already arises in agricultural crops where farmers' preferences for monocultures, plantings of only one genetic type with the highest yield, can result in the ruin of an entire year's crop when a new pathogen appears. Such a situation is what caused the Irish potato famine; the potatoes grown in Ireland were so genetically uniform that they could not resist a new fungus, black rot, when it invaded the crop. A little potato sex—that is, sex among potatoes, not whatever else you might think—would have gone a long way toward keeping the Irish from starvation.

James Thurber wrote a wonderful book titled *Is Sex Necessary?* It may not be necessary for the geckos. But as long as we have pathogens snapping at our genetic heels, it is for us.

CHAPTER 5

WHEN SEX MAKES YOU SICK

While not exactly celebrated in song and story, sexually transmitted diseases are part of our culture, and they are mentioned in a great many works of music, art, and literature. I was taken aback to discover that "The Streets of Laredo," a song I'd dutifully learned in primary school about (I thought) a cowboy lamenting his imminent demise (why he was on the streets rather than decently tucked away in a hospital was never clear to me), has its roots in "The Unfortunate Rake," an eighteenth-century Irish or English folk song about a young man dying of the toxic effects of mercury treatment for syphilis. In alternative versions of the song he is variously also a soldier, a horseman, a lumberjack, and a "bad girl." This thread is taken up with enthusiasm in more modern times, with song lyrics that mention some form of social disease penned by artists ranging from Bernie Taupin to Eminem. AIDS plays a central part in music from opera to rock, as well as in productions such as *Angels in America, Philadelphia, Rent,* and many others not as well known. And of course there is the ever-popular pastime of naming famous people in history who died or suffered from syphilis, which includes, apocryphal or not, Ivan the Terrible, Schopenhauer, King Henry VIII, Rasputin, Napoleon, John Keats, Vincent van Gogh, Adolph Hitler, and Al Capone. The interest is ongoing: A 2004 reexamination of Vladimir Lenin's medical and autopsy records maintains that the Russian leader, who died at 53 with a compendium of ailments, had syphilis as well. While some of these claims may not be accurate, no comparable sport exists in outing famous sufferers of strokes, diabetes, or lung cancer.

Sexually transmitted diseases have, of course, always carried a moral burden, supposedly providing evidence of sinful sex, whether that means adultery, homosexuality, or association with the "wrong" kind of partner. Social critic and writer Barbara Ehrenreich said, perhaps only slightly tongue in cheek, "I was raised the old-fashioned way, with a stern set of moral principles: Never lie, cheat, steal, or knowingly spread a venereal disease." A. L. Baron, in his 1958 book *Man Against Germs,* goes so far as to blame the Russian Revolution on syphilis, long before the Lenin study mentioned above, suggesting that "The orgies of vice in which the Czarina was involved became a public scandal and the peasants of Russia lost their reverence for the nobility. Then, it is said, the germs helped further to raise the red flag of revolution." The Tuskegee syphilis experiment, in which 399 black men were observed untreated for forty years as they suffered through the advanced stages of the disease, many dying of syphilis or its complications, is one of the more appalling episodes in the abuse of the public health system, and part of its horror lies in its link with the intimacy and shame of the disease. The argument over the origin of syphilis, whether it came from the New World to the Old or vice versa, carries overtones of accusation, with the implicit suggestion that the disease is one more symbol of Western imperialism run amok. AIDS alone has so many social and political associations that it has been held up as an icon in the culture wars, and it is on the brink of transforming the African continent.

Sexually transmitted diseases have indubitably changed our lives in a way that other diseases have not. But they have done so not just by providing poster children for or against gay rights, or by making us think about sexuality in the context of a risky era. Diseases transmitted via sex differ from diseases spread through less deliberate contact, through a sneeze or a shared mosquito, in what makes them more likely to spread. The host's most intimate behavior is of paramount importance to an STD's survival. All of this means that some of the ideas we tend to accept about STDs—that monogamy protects against them, for example—turn out to be wrong.

NOT ALL SOCIAL DISEASES ARE CREATED EQUAL

In a not completely outdated euphemism, sexually transmitted diseases used to be called social diseases. This name was coined by Prince Albert Morrow, a dermatologist in New York at the end of the nineteenth century. With a remarkably modern pragmatism, he linked the growing problem of syphilis and gonorrhea in the United States to the need for better sex education. The diseases had strong connotations of sin and secrecy, and physicians were in a quandary about the ethical way to deal with them, particularly among married patients. Should, for example, the wife of an infected man be told about her husband's illness against his wishes? The medical establishment even wondered if people with syphilis should be prevented from marrying at all. Morrow founded the American Society for Sanitary and Moral Prophylaxis in 1905, hoping to educate both doctors and the public about STDs, and he authored the then-classic book *Social Diseases and Marriage,* as well as the rather charmingly titled *Venereal Memoranda.* Although his terminology has fallen out of favor, the social issues associated with STDs remain surprisingly unchanged, and anyone reading the literature on the history of syphilis and society's reaction to it is struck by the parallels to AIDS.

The problem with the term "social disease" is not its coyness, however, but its inaccuracy. In a sense, all infectious diseases are so-cial diseases, because they require their hosts to be in contact, ei-ther directly, as the recipient of a virus-laden droplet passed during a handshake, or indirectly, through exposure to squirming amebae in stream water that has been used by a camper with diarrhea. Fur-thermore, although one can catch a cold from someone during sexual activity, colds are not STDs. What makes a real STD?

A real STD is virtually always transmitted to adults through sexual contact, much to the consternation of all those who fa-vored the doorknob, drinking glass, and toilet seat avenues of in-fection when asked how they acquired their ailments. Many STDs can also be passed from an infected mother to her off-spring, a process called vertical transmission. And finally, any disease involving the genitals or the reproductive tract is not nec-essarily an STD, even if sexual contact can spread it; vaginal yeast infections, for example, can occur in women spontaneously, if the natural balance of microscopic flora and fauna in the reproduc-tive system is disturbed, and while an infected woman can trans-mit the disease to a sexual partner, sex is not the primary mode of transmission. This distinction is important, because as I pointed out in an earlier chapter, mode of transmission drives many things about a disease, including whether it is lethal or benign and how it will spread.

Scientists distinguish between STDs and what they call OIDs, or ordinary infectious diseases, a term I have always rather liked because of its implied conferral of a certain kind of snob appeal to the other-than-ordinary diseases, a class of ail-ments not usually given much glamour. The distinction is not ab-solute, both because one can acquire an OID through sex, like the aforementioned cold, and because occasionally some STDs may be transferred during nonsexual activities, but STDs never-theless share certain qualities that give them their entry into the club. Perhaps most importantly, STDs are rarely if ever immediately lethal. Instead, they tend to decrease the ability of their hosts to reproduce, not by making them mate less but by damaging organs such as the testes and ovaries. From the point of view of the STD, of course, reducing the host's investment of energy into host offspring is beneficial, because more is left over for the parasite to use for its own needs.

Despite their association with wanton behavior, STDs are more conservative than OIDs in whom they infect, at least in terms of the variety of hosts they can live inside. Many OIDs are easily transferred from animal to human, or from one animal species to another, which is why the advent of agriculture and its accompanying domestication of animals was such a boon to disease transmission. For obvious reasons, however, STDs stay within a single host species unless, as is true for some plants and a few kinds of animals, hybridization, or mating between two species, is commonplace. This loyalty, while perhaps flattering, is less than ideal from the standpoint of fighting the disease, because it means that the pathogen need not compromise its adaptation to one host, and can relinquish being a jack-of-all-trades to be master of one. Like a predator that eats only one kind of prey, the specialist parasite can evolve fine-tuned characteristics that allow it to exploit subtleties of its host. Such specialized weapons can both change more quickly in response to the host's counterdefenses and can afford to become targeted enough to find every tiny chink in the host's armor.

From the standpoint of the parasite itself, specificity combined with sexual transmission means that the host must have sex, preferably with multiple partners, or the parasite will die. Simply sitting around in the intestinal tract and waiting for a careless bout of hand-washing after the host uses the toilet, or swimming into respiratory droplets to be sprayed hither and yon with a casual sneeze, isn't good enough. This means that anything the disease does that encourages licentiousness should be favored. Finally, if you are a disease that occurs in a nonsocial host species, or a host that lives at low population densities, you do better if you are transmitted via sex than through other forms of contact, because the one thing to count on in a sexually reproducing animal is that eventually, boy will meet girl, or at least be motivated to try and find her. From the pathogen's perspective, then, sex is just another way to get around.

Also unlike most OIDs, the STD club is adults-only, except under unusual circumstances, and this too has some unique consequences for these diseases. Diseases that can spread to child and adult alike can show wild fluctuations in the numbers of infected individuals, and the resulting epidemics may sweep through an entire population, leaving behind only those whose immune systems could resist the attack. The disease may virtually disappear, only to reappear when infected visitors come into the population. But because STDs can only spread from adult to adult, these oscillations in the proportion of infected individuals will be diminished. In addition, one can never develop childhood immunity to an STD; diseases like measles, mumps, and whooping cough are called childhood diseases because people tend to be exposed to them while young. Because of the nature of children's immune systems, the symptoms are less severe in children, and then immunity for life is conferred. No such opportunity for early resistance exists for STDs, and so they encounter an adult body with full-blown fury.

MITES, WORMS, AND VENEREAL ORF: STDS IN NONHUMANS

Far from being unique to humans, like flush toilets and taxes, STDs are widespread among animals, both vertebrate and invertebrate, and even some plants. A lovely white-flowered weed called white cockle or white campion is subject to the marvelously named anther smut, a disease spread by the insects that carry pollen, the plant version of sperm, from flower to flower. The idea that ladybugs, koala bears, and flowers can get sick from sex puts a new light on the idea that STDs are punishment for sins of the flesh. Like other diseases, STDs are part of how we evolved, but they also have a niche of their own.

The biodiversity of both the kinds of organisms that cause STDs and the hosts that suffer from them is astonishing, though admittedly lacking in the kind of exotic glamour that the same term has when applied to coral reefs or tropical rain forests. Nevertheless, insects, birds, mammals, reptiles, and even snails and toads get STDs, and it is likely that far more exist than are currently known or described. Most of the nonhuman STDs that have been studied occur, not surprisingly, in domesticated species, so that the STDs of pigs, cows, and horses have been reasonably well documented while those of lions, eagles, and fireflies are virtually unknown. Similarly, in keeping with this economic motivation for study, the only species of snail known to suffer from a venereal infection is *Helix aspersa,* a member of the genus eaten as escargots, and the disease itself, a nematode or roundworm infection, was discovered by researchers in France. (Incidentally, if you find the slang for human STDs like "the clap" to be inelegant, be grateful we do not suffer from venereal orf, as do goats.) The few scientists who have tried to systematically survey STDs in nonhumans uniformly caution that the prevalence of these diseases is unknown, but almost certainly vastly underestimated. Hence the impact of STDs on the ecology and behavior of their hosts is likewise still a mystery.

We tend to think of STDs as being microscopic creatures, like the virus that causes AIDS or the bacteria responsible for syphilis or gonorrhea, but larger parasites such as worms can be transmitted sexually as well, and are the primary type of sexually transmitted pathogen in invertebrates. Some ectoparasites, like pubic lice or mites, are also passed from one sexual partner to another. Again, some of these external creepy-crawlies use sex as one of several means to jump from one host to another, while others transfer to a new home only when the hosts copulate. Probably the best known of the latter to constitute a true STD in an animal is a mite that infests European ladybird beetles, familiarly known as ladybugs. The same species of ladybird occurs in England, but only those beetles on the European continent are infected; make of this what you will. Mites are minuscule relatives of spiders and ticks, and they settle under the wings of the beetles and use their straw-like mouthparts to suck blood from their host. The adult female mite lays eggs that hatch into mobile larvae, and when the ladybird beetles mate, the larvae crawl from one member of the pair to the other. Because the beetles are not sociable animals, copulation is pretty much the only contact they will ever have, and so the mites depend on sexual activity for dispersal, making them a true sexually transmitted pathogen. The blood-sucking is more than a minor nuisance for the beetles and in just a couple of weeks can lead to infertility and decreased likelihood of surviving the cold winters.

Several species of worms also creep from the genitals of one sex to the other during copulation, but interestingly (and perhaps thankfully) these too are confined to insects, as are virtually all of the larger multicellular sexually transmitted parasites. Why should this be the case? Why should most of the STDs of vertebrates be bacteria or viruses, while insects suffer from venereal mites and worms? The answer may lie in a surprising fact of insect sex. Human males who wish they could, um, last longer during intercourse may be daunted to learn that for many insects, copulation routinely lasts for hours or even days. Stink bugs mate for over 10 hours; the more gracefully named golden egg bugs copulate for up to two days; and the apparent champion, a type of stick insect, has been clocked at seventy-nine days. Even the lowly flea manages half an hour or so, and what is more, can mate with its nether regions while its mouthparts remain attached to its host, an admirable example of multitasking. The achievement becomes even more impressive when one considers the relatively short lifespan of most insects; if you only live for three weeks, spending several days engaged in sex is a substantial investment of time. The prolonged coupling does not seem to arise from sheer hedonism, however; it may serve to remove other males' sperm from the female, or to keep her from mating with additional males, and variation in the length of copulation both within and among species is a hot topic of study among biologists. From the perspective of a parasite such as a worm, which has to wriggle its way from one individual to another and cannot be passively carried in the ejaculate like bacteria or viruses, the leisurely coupling is a boon, because it allows the parasite ample travel time. And so, scientists speculate, invertebrates have a niche for a type of pathogen not usually able to be transmitted sexually. It is noteworthy in this regard that pubic lice, the lone human exception to the rule, can survive elsewhere on the human body, and do not have to be transmitted during intercourse itself. Whether the freedom from worry about venereal worms is enough of a consolation to those desiring greater sexual endurance is a matter of opinion.

Regardless of their type, however, STDs of nonhuman animals can do a great deal of harm to their hosts, even though they do not kill them outright. The long evolutionary history of STDs in animals, and the concomitant efforts to fight them, may explain a variety of characteristics of animal biology, from how quickly newcomers are welcomed into a social group to the materials used to construct nests. And of course, that perennial conundrum: why most birds lack a penis.

This last point may require some elaboration. Three percent of birds, including ducks, ostriches, and the Vasa parrot of Madagascar, have a penis-like intromittent organ. The other 97 percent mate through contact of the cloaca, a chamber into which the reproductive, urinary, and digestive systems all empty. Aptly named after the Roman word for sewer, a cloaca is also present in amphibians, reptiles, and a few mammals, and in most birds sperm are transferred from the male to the female in an extremely brief process dubbed the "cloacal kiss" by a fanciful naturalist. Anyone still feeling disheartened by the sexual prowess of the stink bugs might take comfort from the blink-and-you'll-miss-it rapidity of mating in most birds. The male simply plops the ejaculate unceremoniously into the cloaca of the female, though some modifications of the male outer tissues exist in a few species.

Scientists have fretted over the evolutionary explanation of the absence of an intromittent organ in birds for some time, with a variety of hypotheses being proposed. They haven't found definitive answers, in part because it is difficult to make and test predictions about the general advantages of a trait when it is present in so few species. But one of the more intriguing ideas is that birds lost the organ back in the mists of time because, given the free and easy exchange of fluids and other unsavory secretions from the gastrointestinal tract in the cloaca, the furtive cloacal contact might have reduced the chance of the transfer of nasty microorganisms. Jim Briskie and Bob Montgomerie, the two ornithologists who proffered this suggestion, point out that it is impossible to evaluate more thoroughly because so little is known about STDs in wild birds. Chickens certainly suffer from

STDs, so there is no reason to think their wild counterparts would be spared, but detailed information is simply not available. I find their idea more convincing than some of the alternatives, including the hypothesis that birds lost a phallus-like organ because of the risk of the male toppling over while standing on the back of the female, a common mating posture in birds; the idea, presumably, is that if mating is rapid, with no need for insertion of the male member, the poor guy keeps his balance better. It seems to me that selection for more gymnastically inclined mates, or perhaps broader-backed females, would take care of this problem more easily, but in all fairness, the jury is still out on this issue as well.

RISKY BUSINESS

If disease is spread through sex, a logical conclusion is that having sex less often, or with fewer partners, will reduce the chance of getting sick. This idea recurs in sociological and epidemiological studies alike, and is a cornerstone of some social policy. It has also been a player in the sex education debate, with some arguing that if you teach children about STDs and the accompanying use of condoms and other means to prevent infection, they will behave more responsibly, and others suggesting the opposite. What does the science say? From an evolutionary perspective, there is only one issue: Will a particular trait increase or decrease the likelihood that its bearer will leave genes, in the form of offspring, in future generations? If that likelihood is increased, the trait, regardless of its temporary ill effects, will become more common. If it is decreased, the trait will eventually disappear.

STDs spread most effectively when an individual has a high number of sexual partners, and several scientists have suggested that some aspects of mating behavior in animals, such as the tendency to avoid new mates until the strangers have spent time in a social group, evolved in response to the dangers of STDs. Here is where animals provide an experimental testing ground for ideas that humans cannot: We can study them to ask, for example, whether monogamous species tend to show patterns of infection that polygamous species do not. Furthermore, information about who mates with whom is sometimes a bit easier to obtain from an observation post in the field than from surveys of humans who may or may not be reporting the truth.

Like many animals, primate species differ in their mating system, with some, like gibbons, being monogamous and others polygamous, like the Barbary macaque, in which a female may mate with up to 10 different males in a day. Charlie Nunn, a biologist at the Max Planck Institute for Evolutionary Anthropology in Germany, wanted to see if the increased likelihood of multiple sexual partners also increased the risk of disease. Nunn and his colleagues examined primates in wildlife parks, not for the incidence of disease itself, but for a signpost of it, the white blood cell count. White blood cells are important immune system components, and in a healthy animal, all else being equal, a high count reflects an evolutionary history of dealing with disease. It is a bit like concluding that people who usually have a hammer on them must frequently need to pound nails.

Nunn and his coworkers looked at the variation in baseline white blood cell counts in 100 species of primates, trying to see whether the species' mating patterns were related to this aspect of their physiology. They found that when females mated with more males, the species had higher cell counts, which means that more vigilant immunity evolves where more sex partners are possible. This correlation might appear because of a factor like group size, with larger groups increasing the risk of all diseases, or like the body size of the animals, with bigger bodies perhaps having higher functioning immune systems. But none of the alternatives Nunn tested were associated with the white cell counts.

If STDs are harmful, and having more sex partners—at least infected ones—increases the likelihood of getting one, it is logical to conclude that natural selection should act on hosts to reduce the risk of transmission by affecting the way the hosts mate. This might occur at two levels. First, it is reasonable to imagine that a high prevalence of STDs would make animals—and people—less promiscuous, so that when STDs present a heightened or new danger, monogamy, or at least fewer sexual partners, would be favored. Even if a species has evolved the enhanced immune defenses apparent in Nunn's survey, from an individual perspective it makes sense to minimize the need for such defenses. Second, animals should attempt to discriminate among potential mates and refuse those that appear to be infected. It turns out, however, that neither of these seemingly sensible suggestions is borne out in nature, and in fact, from a theoretical perspective this is not surprising. To understand why, we need to turn our attention once again to the parasite's point of view.

Should an individual be monogamous? As with all other traits subject to selection, the answer is: only if fidelity helps spread the bearer's genes. And that is often not the case, at least in nonhuman animals. For males in particular, mating with many females increases the number of offspring they sire, regardless of whether the male contracts a disease that eventually reduces his fertility. Females may also benefit from mating with multiple males, including gaining access to resources such as food, protection of their young, and greater chance of a truly high-quality father for at least some of their offspring. Short-term gains can offset long-term losses, so while STD infections increase with the number of sexual partners, individuals who reduce their number of partners are simply outreproduced by those who do not. Even when most members of a population have only one or two sexual partners, mathematical models suggest that a few highly promiscuous individuals, called "superspreaders," can ensure a relatively high infection rate in the group as a whole, rendering the efforts of the chaste masses moot.

British scientists Michael Boots and Rob Knell used complicated mathematical modeling to determine that individuals with two different approaches to risky sex could coexist in the same population and still pass on their genes. One type was indeed highly risk-averse, the chaste and cautious individual that many would have predicted to be the product of a world rife with STDs. But equally successful, it turned out, was an extraordinarily reckless type. STDs didn't select for monogamy, they selected for variability. Monogamy works, to a point, but so does extreme promiscuity. From the disease's perspective, of course, it doesn't matter whether it is spread quickly by a few individuals or slowly by many, so long as it persists.

An enormous complication in trying to apply these results to modern human populations is our ability to have our cake and eat it too, so that sex with condoms or other protection can eliminate the effects of STDs without eliminating the sex. This can influence not only sexual behavior of the host but the virulence of the disease. For the moment, however, the point is that monogamy is not the only effective defense against STDs.

What about mate choice? Even if the threat of STDs doesn't change with the number of partners, why wouldn't it pay to choose only uninfected individuals and shun the rest? The immediate answer is that it would, but as with the monogamy, the costs of prudence may outweigh its benefits. Several studies of STDs in nonhumans have attempted to see whether infected males are chosen as mates less frequently than their healthier counterparts. One of the most thorough used the ladybirds subject to the blood-sucking mites described previously. Mary Webberley and her coworkers first surveyed the beetles in the field; mated pairs remain coupled for some time, like the other insects mentioned earlier in this chapter, and she simply wanted to see if uninfected individuals would be found paired up more often than would be expected by chance. If they were, it would suggest that healthier insects were more likely to mate. This is not a definitive answer to the question of whether infected mates were rejected, since other factors might explain the disparity, but certainly encourages further research. But it turned out that mated ladybirds were just as likely to have the sexually transmitted mites as single ones. Webberley then went on to place infected and uninfected insects in a small dish, a ladybird version of a singles bar, and recorded their behavior. Despite her cheerful appearance, a female ladybird that doesn't want to mate can indicate her disdain quite forcefully, and she is quite capable of exerting choice based on a number of characteristics of a potential suitor. Again, however, and somewhat to the researchers' surprise, female ladybirds were just as likely to reject a healthy male as a mite-ridden one. Few other studies like this have been done, but all seem to suggest the same conclusion. Why animals are not more discriminating remains a mystery; perhaps the risk of missing a chance to mate at all is simply too great.

Rather than looking at mate choice directly, Charlie Nunn turned his attention to ways primates might avoid becoming infected with STDs via their social behavior. He sent a questionnaire to researchers who study primate behavior and ecology, asking them if they had observed behaviors that might reduce the acquisition of pathogens during sex, such as genital examination of partners, grooming the genitals, or urinating immediately after mating. This is not as oddball a request as it might sound, incidentally; many primate ecologists make detailed observations of their charges and are aware of even their most intimate activities. And many animals, not just primates, show distinctive displays or postures immediately after mating. Hens, for example, always follow copulation with a brief vigorous shaking of their feathers. I have always wondered if this functioned to dislodge any mites or lice that had crept from the cloaca of the male to the female. Nevertheless, even though it seems logical for animals to reduce the likelihood of contracting an STD, no evidence from Nunn's surveys suggested that they did.

What's going on here? The short answer is that having sex, even risky sex, is almost always worthwhile from an evolutionary perspective, perhaps a daunting thought if one is working in public health but a biological truism nonetheless. This is partly because of the short term/long term dilemma, and partly because of a rather cruel twist of fate: Highly attractive males, those that are preferred by a great many females, are also the most likely to be infected, simply because they have been doing more mating. If the choice is between a loser without an STD and a superstar who might or might not have one, females are still likely to benefit by choosing the latter. Furthermore, it may be more difficult to detect a suitor infected with an STD, as opposed to one that is diseased in general, than one might like. From the parasite's perspective, it is best not to have too obvious an effect on one's host, since anything deterring potential mates also reduces transmission of the disease. So it should pay for the pathogen to become cryptic, at least in the early stages of infection, further circumventing efforts of sexual partners at avoiding infection. Indeed, many human STDs show few or no signs for weeks after infection, and sometimes not for months and years. All of these factors conspire to make mate choice on the basis of STD infection an iffy proposition at best.

Females, particularly mammals, are at an STD disadvantage

for another reason; an infected male is more likely to infect a female than the reverse. This follows from the mechanics of sex; the male's ejaculate remains in the female long after intercourse, and hence the disease-causing organisms have a longer opportunity to invade the female's body. In addition, the delicate mucosal surface of the female reproductive tract is much more subject to microscopic scrapes and tears than the penis, allowing further assault by pathogens. Interestingly, female mammals also have high levels of immune components called immunoglobulins in the lining of the reproductive tract, and after sex, phagocytes, a type of white blood cell that generally engulfs and breaks apart foreign objects in the body, congregate around the ejaculate. This asymmetry between the sexes supports the idea that we show a cellular signature of past encounters with STDs, and that they are hardly a creation of modern immorality.

AN UNUSUAL SORT OF NIGHTCLUB

Going clubbing turns out to have more to do with STDs than OIDs for more reasons than just the obvious one. STDs depend on the frequency, but not the density, of infected individuals in a population, whereas the reverse is true for OIDs. To explain this distinction, Ann Lockhart, Peter Thrall, and Janis Antonovics, biologists who originally got interested in STDs because of their work on the anther smut disease of the wildflower, use the analogy of people at a nightclub. Imagine that you are dancing with other individuals at the club more or less at random, or at least that the chance of your being within a few feet of any particular person is equal by the time the club closes. At the end of the evening, let's say that you go home with one person and have sex with him or her.

While this scenario raises any number of fascinating social

and personal dilemmas, let us focus on disease transmission. Your chances of catching a cold or other OID after your night's activities depend on the number of people in the club, since you'll be within sneezing range of most of them before the night is over. Assuming the fraction of people in a nightclub with colds is the same regardless of how big the club is, your own likelihood of encountering the virus increases with the number of people there, or with what is called the density of the population. You're more likely to sniffle after being in a club of 400 than one of 40.

Not so for the STD, unless, as the authors of the study say, "it is an unusual sort of nightclub." If you only have one sexual partner from the crowd, your chance of getting an STD depends entirely on the likelihood that this one person is infected, regardless of the size of that crowd. And that likelihood in turn reflects the proportion of people in the club with the disease, not the total number of people in the club. If one out of ten people is infected, your chance of engaging in your tryst with a healthy partner is 90 percent, regardless of whether there are 40 or 400 other people in the club at the time. If half of all the people in the club are infected, however, your chance of finding an uninfected partner is likewise 50 percent, again independent of the total size of the club.

Now, clearly these extremes are exaggerated. No one truly encounters everyone at a nightclub, or within a population, at random, and of course some sexual partners are more likely to have an STD than others. And there are many unusual nightclubs. But the general principle seems to hold true for many STDs, and the consequences are important. For example, an OID is often controlled through vaccinating a large proportion of a population. Once only a few individuals have the disease, it won't spread because the infected individuals recover or die before transmitting it. However, even a few contagious individuals can spread an STD fairly rapidly. Vaccination against an STD still has the potential to curb an epidemic, but the details for achieving eradication are more complicated than they are for an OID, partly because of the age at which vaccination has to occur to be effective. Add to this the tendency for STDs to mask their symptoms early in the infection, and for the disease to linger in those infected for much of their lifetimes, and you have a daunting problem on your hands.

WHY SMALLPOX IS SMALL: STDS AND THE FUTURE

The AIDS epidemic has led many people to speculate about whether this deadly disease will continue to ravage the planet, or whether it will become more benign. We know that the conventional wisdom that all diseases proceed to a less severe state is untrue, but at least some do, and the question is which ones, and why. What about AIDS, or other STDs? Will AIDS eventually become a nuisance, like pubic lice, or will it continue to wreak havoc on the world?

To put things in perspective, it is useful to consider an older STD than AIDS, namely syphilis. Detailed records of syphilis infection start appearing in Europe in 1495, and a fearsome disease it was. Smallpox was called smallpox to distinguish it from the Great Pox, syphilis, which caused, in the words of Ulrich von Hutten in 1519, "Boils that stood out like Acorns, from whence issued such filthy stinking Matter, that whosoever came within the Scent, believed himself infected. The Colour of these was of a dark Green and the very Aspect as shocking as the pain itself, which yet was as if the Sick had laid upon a fire."

Two points are noteworthy about this vivid description. First, it contrasts markedly with modern experiences with the disease; although serious in its overall effects, the rash and other overt symptoms of syphilis are now much more muted, and the disease may go undetected for some time. Second, it is reasonable to suppose that a sufferer of the symptoms von Hutten describes would be unlikely to get a lot of dates.

These two observations led Rob Knell, one of the scientists working on the ladybugs' mites, to propose that they were connected. In other words, if a syphilis-ridden individual were less likely to have sex, and hence spread the disease, the disease organism would do well to evolve a less acute effect on its hosts. Syphilis, he argued, became less severe because it was transmitted more readily if victims were still attractive to the opposite sex, or at least still interested in pursuing romance. Note that although several symptoms of the disease have changed, the most pronounced difference is in the presence of those pustules on the skin, which are perhaps the most graphic signal of infection to prospective mates. Knell pointed out that the host couldn't have evolved resistance so rapidly, but the syphilis bacteria certainly could, since even a few years represents many thousands of generations. So we have syphilis itself to thank for the lessening of its symptoms. (Of course, the disease is still serious; its symptoms may have ameliorated, but the prognosis in untreated victims remains grim.)

Can we expect a similar reduction in AIDS virulence as time passes? Best guesses are not optimistic. Virulence is linked with transmission rate. On the one hand, HIV should benefit from keeping its hosts alive (and having sex with more people) for increasingly long spans, just as the syphilis bacterium does. And HIV in most parts of the world disproportionately affects subpopulations, such as intravenous drug users, sex workers and their patrons, and gay men, along with the sex partners of all of the preceding. This structuring means that an epidemic will not continue indefinitely; the host populations will become saturated, reducing the fraction of people who become newly infected and hence stopping the spread of the disease.

Bruce Levin, James Bull, and Frank Stewart, American evolutionary biologists interested in mapping the spread of emerging diseases, suggested that the number of new HIV infections in one of these subgroups might go down not due to efforts at education or the use of condoms, but because of this saturation of the available host population. Treating AIDS with drugs will curb the epidemic if the therapy not only makes people feel better but also reduces the likelihood of their spreading the virus to others by, for example, lowering the numbers of virus-infected cells in their bodies. The more people who get a pathogen, the greater the opportunity for selection. It might seem that in the interim, selection for those resistant to the effects of the disease should eventually result in a human population that does not suffer from AIDS, or at least manages it in the way that we manage a common cold. While this is possible, the biologists believe that such an adaptation would require thousands of years to take effect, and that hordes of people would die from the disease before it happens.

Will the virus itself evolve, as may have been the case with the syphilis bacteria? Here the situation is complicated, and different experts make different predictions. Remember that HIV infection already shows few symptoms in the early stages, unlike the fifteenth-century version of syphilis, so it doesn't face the same evolutionary pressure to become less virulent. Paul Ewald, one of the champions of Darwinian medicine, has suggested that as with other diseases, easier transmission should select for increased virulence of HIV. A change in transmission, as might occur via a reduced number of sexual partners, should then select for a more benign pathogen that doesn't harm its host as it waits for its chance to go to a new victim. Although plausible, this scenario has been questioned by Levin, among others, who says that a great deal depends on whether the disease is epidemic and spreading or endemic and holding steady. If being virulent helps a disease transmit early in its spread through a population, then it will do well during the epidemic phase even if it kills its host at a relatively early age. But if the disease is well established in a population, it's more important from the pathogen's point of view to hang in there and capitalize on any new hosts that come along, which may select for a relatively benign illness. Wouldn't the ideal STD not only be cryptic, but actually enhance its host's sexual urges, or perhaps make that host even more irresistible? In the movie *Shivers,* a parasite that was originally supposed to repair people's defective body parts turns on its creator (don't they always) and instead makes its hosts seek out sex with anyone they can find so that it can be spread. The bearer of the parasite is not necessarily more attractive, but it is certainly more voracious.

There is no direct evidence of a parasite doing anything like this in animals, much less humans, but that does not mean such a pathogen doesn't exist. It would, in my opinion, be worth examining the effect of STDs on the sexual proclivities of their hosts. An insect close to home, however, comes close to illustrating a similar ploy. Ordinary houseflies are subject to infection by a fungus that invades their abdomens and expands to fill the body cavity. Eventually the parasite kills its host, but the body remains glued to a leaf or grass stalk by the strands of emerging fungus. The spores from the fungus then infect other flies that come into contact with the victim. None of this would involve sex except for one thing: The fungus makes the abdomens of infected flies look swollen, balloon-like. And there is nothing more attractive to a male fly than a female with a large abdomen, because that means she is full of eggs that he can fertilize. Anders Pape Møller, a behavioral ecologist in Paris, had the ingenious idea of giving flies an opportunity to approach dead flies with and without the infection. Male flies were much more attracted to dead flies with the fungus than to uninfected corpses, and once attracted they often tried to mate with the body, exposing themselves to the parasite. Whether the swelling of the abdomen is actually an adaptation that lures a new host by exploiting its urge to procreate, or a happy side effect of infection, the end result is the same: a parasite that spreads by making its host more sexually attractive. That the host is also dead at the time makes no difference to the pathogen.

GOOD BACTERIA AND A RAY OF HOPE

If all this talk about sex causing illness is too disheartening, a proposal by Michigan ornithologist Mike Lombardo might cheer you. He wondered why, in the face of all the risks associated with sex, including STDs, female birds nevertheless seemed to mate repeatedly, either with the same male or several different ones. In many species, a single copulation is sufficient to fertilize all a female's eggs, but she will continue to mate long after she has enough sperm. Lombardo suggested that the cloacal contact between the sexes transferred beneficial microbes that may yield protection against more nasty bugs. The idea that some bacteria and other microorganisms are actually helpful is not new, of course, as the ads for yogurt and other health food products show. The normal gastrointestinal tract of most animals, whether bird, reptile, or human, is teeming with microscopic fauna and flora. Some animals, like termites, even require an inoculation of a particular type of microorganism so that they can digest their food properly; the termites get a careful dose of microbes from a nestmate that allows them to deal with the otherwise completely unusable cellulose in the wood they consume. So it is not altogether

CHAPTER 6

THE SICKER SEX

My father went to live in a nursing home as he approached 80, and at least initially he was able to enjoy the doting attention of the other residents. Once when I visited him an aide said how much his presence was appreciated at the dances the facility held from time to time. "He's a great dancer," the aide enthused. While happy to hear that my dad was settling in, I was a bit taken aback at the praise, since to my knowledge my father had never danced before in his life. Then I realized they didn't appreciate his dancing per se but his mere presence: For the crowd at the home, being a man was all it took to make him a virtual Fred Astaire. Well, being able to stand up was also helpful, but the paucity of male partners guaranteed my father's popularity.

The female-biased sex ratio in facilities and organizations for the elderly is not news; most people realize that women generally live longer than men and widows outnumber widowers. What is not so widely appreciated is that the female bias holds at all ages, regardless of the cause of death. Arguments about the weaker sex notwithstanding, there is no contest about the identity of the sicker sex. It is males, almost every time. People will tell you it's because of smoking, it's because of heart disease and the protective effects of estrogen, it's because we encourage adolescent boys to take foolish risks, but the truth is that none of these causes can fully explain the disparity.

Smoking and teenage motorcycle crashes certainly cannot

explain the same pattern of male vulnerability in a group of small marsupials called antechinus, rodent-like creatures that live in Australia. They are sometimes called marsupial mice, but they are more closely related to the kangaroos and other pouched mammals than to real rodents, though you would never think so when looking at them. There are several species of antechinus, but all have a breeding season lasting only a few weeks, timed so that the females will be nursing their young when the insects and other invertebrates the antechinus eat are at their spring and summer peak abundance. They are generally solitary except for the breeding period, when males (which in the brown antechinus are twice the size of females) become completely obsessed with sexual activity. All of the animals in a population are ready to mate at the same time, which means that males spend the rut frantically competing with other males for access to the sexually receptive females. During this single-minded pursuit of romance, the males' digestive system breaks down as feeding is neglected, and levels of corticosteroids, the so-called stress hormones, skyrocket. The hormonal surges eventually trigger a failure in the immune system, so that male antechinus become unable to defend themselves against disease. After the rut is over and the females are pregnant, all of the males die, completely spent, like salmon after spawning. Even more remarkable, females stay alive not only until they have weaned their young but for up to an additional two years, making their longevity triple that of males. In the nursing home, that is the equivalent of octogenarian women surrounded by senile twenty-five-year-old men, or of a few 75-year-old geezers in a room full of women aged two centuries or more.

Why are males so much more vulnerable to disease than females? People are hopeful that we will close the gap between men's and women's lifespans as we develop better medical care and education about health risks, but if the difference evolved in response to disease it is not going away any time soon. It is buried too deeply in our makeup to be cured with a nicotine patch or a course of antibiotics.

BEING MALE IS A HEALTH RISK

The first clue that the difference in male and female longevity is deeply rooted comes from examining the variation in lifespan that occurs in different cultures and at different times in history. A group of researchers from France and Russia surveyed 227 countries, from Afghanistan to Zimbabwe, and found that women outlive men in all but a handful of countries, whether their lifespan is short, as in Sierra Leone (49 years for women, nearly 44 for men), or long, as in Norway (82 years for women, 76 for men). The few countries where men outlive women are almost all in a state of HIV- and conflict-churned crisis, like Zimbabwe, where women live a scant 35 years to men's 38. The chart the scientists published is amazingly compelling reading, particularly for trivia lovers—who knew that women in China, Colombia, and Belarus all have about the same lifespan of about 74 years, whereas their men die at 70, 67, and 62?

Even more arresting is the finding that the gap between male and female longevity increases the longer both sexes live, so that Finnish women live to 81, but Finnish men gasp their last at 74, a difference of 7 years; in Egypt, on the other hand, neither sex lives as long and while women still outlive men, they do so by only 4 years, 66 to 62. Other patterns that cry out for speculation about their cause include an increase in male mortality in Europe and Christian countries over the last several decades, but higher female mortality over the same time span in Asia and in Muslim countries. Could it be that more mothers die in childbirth in the latter? Is alcoholism a factor in the former? Regardless, the consistency of longer female lifespan is overwhelming.

Daniel Kruger and Randy Nesse at the University of Michigan took a slightly different approach. Instead of simply examining overall lifespan, they compared men's and women's mortality rates for eleven causes of death, from accidents to homicide to liver and heart disease to pneumonia, in twenty countries. Again, the women win, and the authors conclude, "Being male is now the single largest demographic risk factor for early mortality in developed countries." It isn't that men tend to die from things that never affect women; a list of the ten leading causes of death in the United States in 2000 shows that heart disease, cancer, and cerebrovascular diseases (such as strokes) are the top three for both sexes. Instead, almost all causes have disproportionate effects on men.

This trend has held remarkably steady over time, too. According to the United States Census Bureau, in 1900 the life expectancy for German men was 44, rising to 65 in 1950 and 74 in 1998. Women's life expectancy rose right alongside, going from 47 in 1900 to 69 in 1950 and over 80 in 1988. In developing countries like Thailand, information wasn't available for lifespan at the turn of the century, but the comparison of 1950 and 1998 yielded the same result, with men going from 45 to 64 and women from 49 to 73. Get as old as you want, boys—the girls just outlast you.

It is important to remember that an average life expectancy of 35 or 40 does not mean that most people get to that age and then die. Average life expectancy is just that, an average of all the ages that the people in the population attain before they die. A life expectancy of less than 40 can occur without a single individual dying at or even near that age, if for example childhood mortality from diseases such as measles or malaria is high, a common pattern in developing countries. Say that in a population of ten individuals, five die at age 5, two at age 60, and the remaining three at 75. The average lifespan in such a hypothetical society is therefore 37, but no one reached their thirties hale and hearty and then suddenly began to senesce. The same pattern writ large is what makes the life expectancy in developing countries so shockingly low. It isn't that people in sub-Saharan Africa or ancient Rome never experienced old age, it's that few of them survived their childhood diseases long enough to do so. Old age is not a recent invention, but its commonness is. So we need to ask not what kills old men at a higher rate than old women, but what kills males of all ages, thus contributing to the sex difference in longevity.

IN SICKNESS AND IN HEALTH

Females don't just live longer than males, they also suffer less from a variety of diseases. People have known this for some time, yet it seems they have trouble coming to grips with it. In a 1958 paper titled "Biological Sex Differences with Special Reference to Disease, Resistance and Longevity," a physician named Landrum Shettles listed numerous ways in which males suffered more from ailments or were simply more fragile than females, including being more sensitive to radiation, and concluded, "Females are more resistant to disease, the stress and strain of life.... In general their biological existence is more efficient, pre-eminent than that of males. In brief, the human male with beard and functioning testes pays the higher price." Am I alone in detecting a note of peevishness here? Perhaps it seems too difficult to reconcile the hardiness of women with the stereotype of the weaker sex.

Scientists have usually focused on one of two kinds of explanations for this pattern: ecological, or what the sexes do that exposes them differently to disease; or physiological, how the internal workings of the sexes make them differently susceptible. The ecological differences were noticed by scientists who pointed out that if men and women, or female and male animals, ate different things or went to different places, they often ended up with different kinds of parasites. For instance, more male than female dogs have heartworm, perhaps because males roam more than females and are used more often in hunting, both behaviors that expose the dogs to the mosquitoes that are the vectors of the worm. In parts of Africa, more boys than girls get schistosomiasis, a disease caused by a parasitic worm that passes from aquatic snails to the bloodstream of humans and eventually to other body organs such as the small intestine or the bladder. At one time blood in the urine of a boy in countries where schistosomiasis was prevalent was viewed as a natural rite of passage akin to menstruation, a sign of maturity rather than an indication of damage by parasites. The worms themselves are indiscriminate about the sex of the bodies they enter, but boys are more exposed because they swim in the ponds and rivers of their villages, urinating in the water and transmitting the eggs as they do so. Girls are not as likely to become infected because they are not permitted the same freedoms.

Even if males and females have the same likelihood of harboring a parasite, the sexes can differ dramatically in how they transmit disease because of behavior. A modest-looking rodent called the yellow-necked mouse hosts a tick that carries viral encephalitis, a nervous system disease that can infect humans as well as the mice. Mature male mice are much more likely to transmit the disease than females or young males, probably because they roam over a wider area in their search for mates.

Many of the things males do to attract females also make them more vulnerable to disease. Spadefoot toads are drab desert dwellers that spend much of the year burrowed deep in the sand, waiting for the brief periods of rainfall when ephemeral pools fill with water. During these windows of moistened activity, males are continually immersed in water, croaking their ardor to passing females. The female toads visit a pond once in a given year to mate, after which they lay their eggs and return to a life of somnolence. The longer a toad spends in a pool, the higher the risk of infection by a wormlike creature called a trematode. Males get more trematodes simply because they expose themselves for a longer period, a sacrifice they are prepared to make to increase their mating opportunity.

In vertebrates, the physiological explanation has its roots in a little-known side effect of the very substance that makes males masculine: testosterone. Men need testosterone to grow beards, build muscle mass, and make sperm. Other male vertebrates use testosterone to produce songs, in the case of birds, or to develop the fleshy comb and wattles of a chicken or turkey, as well as to stimulate energetic courtship displays. In humans and other animals that are able to reproduce year-round, testosterone levels stay relatively stable year-round as well. In animals that have a brief breeding season, testosterone surges in the bloodstream just before the males—and their testes—get ready for mating. In some species, testosterone is also connected to aggressive behavior, so that artificially elevating its levels in the blood makes male mice, for example, get into more fights with other males. Men with higher levels of testosterone were judged as having more "masculine-appearing" faces when subjects saw photographs of composites from the faces of men with high or low levels of the hormone. It's probably fair to say that from the standpoint of most men, testosterone is a good thing, a symbol of virility, something a guy can be proud of.

What is not so widely known is that testosterone also has a

dark side. I am not referring to the stereotypical complaints about men leaving socks on the floor, being endlessly fascinated with televised sports, or communicating via grunting. The truth is that testosterone makes men sick, or at least sicker. More properly, it is associated with a weak immune system and increased susceptibility to disease in males from a wide variety of vertebrate species, including humans. The hormone can depress the ability of other immune cells, tissues, and organs to develop, or it may actually help parasites such as worms grow inside the body. More indirectly, testosterone can cause the levels of other hormones like cortisol, one of the so-called stress hormones, to increase, and cortisol in turn suppresses the immune system further (which is why people can get colds or other ailments more often when they are under stress). Interestingly, men in committed romantic relationships tend to have lower testosterone levels than unattached men, which fits in with the well-known finding that married men live longer and are healthier than unmarried ones. Obviously, many other factors influence the well-being of married versus single people, but it is at least intriguing to speculate that reducing testosterone once a man is paired has the side benefit of improving his immunity.

Among mid-twentieth-century parasitologists, something of a cottage industry developed of infecting male and female experimental animals with parasites before or after removing their testes or ovaries. The scientists almost invariably found that administering worms to both male and female animals resulted in the males harboring a heavier infection. Removal of the testes erased the sex difference, and an injection of testosterone usually caused the male's susceptibility to reappear. I have sometimes wondered what a psychologist would say about why these experiments were so compelling, because they were repeated over and over, with at least a dozen kinds of worms in at least as many kinds of hosts, and almost always, I might add, performed by male scientists.

Females may benefit not only by their lower levels of testosterone but because of the protective effects of their own sex hormones. Estrogen seems to protect women against some forms of heart disease. Women also seem to produce more new T cells, those immune system components that originate in the thymus. British researchers who compared deaths from pneumonia and influenza in men and women found not only that more men than women died as a result of the diseases, but that the difference was reflected in the higher level of thymic activity in the women. Female mice infected with a one-celled parasite produced more antibodies than males infected with the same dose, probably due to hormonal advantages.

Interestingly, the only entire class of diseases that is generally more severe in women than men are autoimmune disorders, like lupus erythematosus, rheumatoid arthritis, and multiple sclerosis, in which the immune system is in effect too vigilant, so that it attacks the cells of the body as if they were nonself. This hyper-reactivity is triggered by a variety of cues, not all of which are well-understood; a too-clean environment in childhood with insufficient immune stimulation is one possibility, as I mentioned in the case of allergies. Women seem to be stuck with overly alert immune systems, and are thus more vulnerable to this overshooting the mark. Pregnancy poses its own immunological challenges, because a mother must not reject her own fetus as if it was a foreign transplant, and yet it contains cells derived from another individual, the father. Pregnancy is often accompanied by a reduced immunity, and women with certain autoimmune diseases like rheumatoid arthritis can experience a temporary relief from their symptoms when they are pregnant.

A few infectious diseases are more common in females than

males. A tapeworm of rodents and dogs called *Taenia crassiceps,* for instance, is more likely to be found in females than males, and curiously, estrogen seems to help the worms grow while testosterone hinders their development. Malaria sometimes seems to be more severe in females and sometimes in males. Even when a difference in parasite levels is male-biased, the gap is rarely huge. Females and males are very much alike, after all, and there are no diseases that infect only one sex. But as with runners in an international competition, small advantages can make all the difference in determining the winner.

The physiological and ecological or sociological causes of the sex difference in susceptibility do not work alone; heightened testosterone levels in a male bird, for example, may induce him to fly more actively in his territory, increasing his exposure to flies or mosquitoes that transmit malaria. At the same time, the suppression of his immune response by the same hormone will also increase his likelihood of suffering from disease. More men than women died in the 1918 influenza epidemic, perhaps because they were also more likely to suffer from tuberculosis, and having TB seems to increase the severity of flu. We don't know why more men had tuberculosis, but it is certainly possible that their lifestyle increased their exposure to the bacteria that cause the disease.

The discovery that males are sicker has important health implications, some practical and some theoretical. The first is simple: If men and women differ in their response to infection, or in their exposure to it in the first place, it makes sense to devise medical treatments based on sex. Doctors are starting to implement separate protocols for the diagnosis and treatment of heart disease, where women are known to experience different symptoms than men, but haven't extended this thinking to treatments of other diseases, particularly infectious ones. Some researchers, such as Jean Woo and Suzanne Ho, both professors of medicine in Hong Kong, argue for a "men's health" movement, similar to the emphasis on women's health that emerged in the 1970s. Such a program would advocate health promotion activities such as quitting smoking and decreasing alcohol intake, since men are more likely to smoke and drink to excess.

My view is that while it is reasonable to try and shore up the male body in all its frailty, it is important not to degenerate into an argument about whose sex deserves more medical attention. A more reasonable approach might be to make sure that we test drug treatments in both sexes, and that when we diagnose disease, we routinely take sex of the patient into consideration. This may be especially important in developing nations where parasitic diseases such as intestinal worms that differ in the sexes are more common and where medical resources are particularly limited.

LIVING HARDER, DYING YOUNGER

We already know that short-term gains have the evolutionary advantage over long-term investments, so dying young might be expected under certain selective regimes. And if males need ornaments, and ornaments require testosterone, they are the unlucky sex. But one question still remains: Why are males the sex with the ornaments, and females the sex that gets to choose among an array of displaying suitors?

Understanding this twist of fate requires understanding sexual selection, the process similar to natural selection that produces traits useful in matters dealing with sex rather than survival. Although evolution is always about leaving the most copies of one's genes in succeeding generations, the sexes go about achieving this goal in different ways.

Imagine a female animal, say a mammal like the brown antechinus I mentioned earlier. She can have between four and twelve young in a litter, and at most she lives for three breeding seasons, so her reproductive jackpot would be thirty-six young, a staggering number rarely if ever realized. Most females will have far fewer offspring, but none will ever lack for a male to inseminate them. They are limited by the number of babies they can produce and rear, not by the number of males that are attracted to them. A male antechinus, on the other hand, if he is successful at wooing females and deterring rivals, could realistically manage to inseminate far more than three females. It is hard to calculate the maximum number of young he could sire, but it is certain to be more than thirty-six. In his case, however, the prospect of a big win is balanced by the greater likelihood of a catastrophic loss in evolutionary terms, namely, no mating success at all. If one male in a population monopolizes ten females, that's nine males left twisting in the wind, paternally speaking. So unlike the females, males are limited not by the number of offspring they produce, but by the number of females they can mate with, which is tied to the number of competitors they can fend off. Males are thus far more variable in the number of offspring they can have, which means that they are playing a high-risk, high-stakes game. Hence the elaborate ornaments, huge weaponry like antlers on elk, and the otherwise pugilistic nature of many males during the breeding season. Selection acts differently on males and females because different pressures limit their reproductive output.

As we have all noticed, however, males differ in just how much machismo they exhibit, and this variation occurs among species as well as within them. A male elephant seal is the equivalent of Sean Connery as 007, a gibbon male more like Tom Hanks, not in physical appearance (though that might bear further study), but in the degree to which competing with other males rules their lives. For the elephant seal, life is all about keeping other males away from a group of females he has been able to sequester in a tiny patch of beach; males are two to three times as large as females and fight ferociously, with a high risk of life-threatening wounds, to maintain access to their mates. The stakes are high: A single male is capable of fertilizing up to 90 percent of the females in a given population, but a female either will or won't have a single pup. Most males cannot even enter the game, and garner no mates at all.

The gibbon, a long-limbed ape native to the forests of Asia, lives a life that should be the pride of every monogamy-loving neoconservative. The male and female stay paired for an extended period and rear their offspring cooperatively, defending their territory together against intruders. For male gibbons, unlike the elephant seals, acquiring more mates will not increase mating success, since a hefty contribution from both parents is required to realize any return on their investment at all. Beating up other males is therefore not a productive pastime. Snow geese, penguins, a few kinds of rodents, and even some types of cockroaches are similarly faithful to their partners, sometimes for life. Other species fall somewhere in the middle: Males compete for mates, and may be able to fertilize more than one female, but lack the boom-or-bust sexual economy of the elephant seal or the antechinus. The usual rule is that the more different the sexes look, the greater the likelihood that one sex or the other engages in ruthless sexual competition, and hence the more likely the species is to have an elephant seal type of mating pattern. Selection for good fighters has led to massive size differences between males and females in animals like the antechinus and seals, whereas both sexes in gibbons or penguins look pretty much the same once they become adult. Humans are the subject of some controversy in this regard, because we still have a fair amount of this sexual dimorphism, as it's called (not as much as if we were like the elephant seals, in which case men would weigh an average of two or three hundred pounds to women's hundred and twenty-five), but we also stress monogamy in most societies, at least in word if not always in deed.

Species therefore differ in how much the males compete for mates, with the more polygamous ones favoring males with high testosterone levels and go-for-broke investments in machismo. If these attributes are also accompanied by costly vulnerability to pathogens, we expect monogamous species to show a smaller difference between the sexes in disease susceptibility than polygamous species. Males may generally be the sicker sex, but how much sicker would therefore depend on the mating system of the species. So male and female gibbons ought to be similar to each other not just in appearance, but in their disease resistance, while male and female elephant seals should differ not only in body size but in their parasite susceptibility. In other words, where the payoff is highest, you ought to see gambling with the highest stakes.

Testing this idea is not as straightforward as it might sound. Looking only at a single species or even handful of species might not yield reliable results because some other factor could explain any difference between the sexes in disease susceptibility. Furthermore, we do not know the details of mating behavior for many if not most kinds of animals. Sarah Moore and Ken Wilson from England found an ingenious way around the problem by using a simple surrogate for the degree of male reproductive competition: the size differential between the sexes. Species in which the males are relatively larger are presumed to have experienced more male sexual competition in their evolutionary history. They used information available in the scientific literature on body size and parasites from 106 different mammals ranging from deer to elephants to mice, and they also used sophisticated methods for taking into account the ancestral relationships between species. The parasites included single-celled organisms, worms, mites, fleas, and ticks. Viral and bacterial infections, though potentially extremely important in the lives of the animals, are simply too difficult to document in wild populations, so Moore and Wilson left them out of the analysis.

Moore and Wilson first found the same pattern that had been established in smaller studies: Males had persistently higher levels of parasitism than females. As they predicted, the greater the difference in size between males and females, the greater the disparity between male and female parasite levels. The scientists also gathered data on longevity in the sexes, and found that when males died at a younger age, they also had a disproportionately higher level of disease. Furthermore, where they could at least classify species into those in which males had the potential to mate with more than one female and those that were monogamous, Moore and Wilson showed that male-biased parasitism was more likely in the former. This is consistent with the idea that intense male competition leads to males being the sicker sex.

A similarly heroic survey was undertaken by András Liker and Tamás Székely, Hungarian scientists now based in the UK, who compared mortality rates in 194 species of birds. Birds have some odd differences from mammals, including a departure from the pattern of males dying younger; most birds show the opposite pattern. This reversal has traditionally been attributed by biologists to the high cost, evolutionarily speaking, of females having to lay the eggs, keep them warm, and bring them worms, often without much help from the male. At the same time, birds also show a lot of the same male competition seen in mammals, so Liker and Székely wanted to see if the difference between male and female longevity grew larger with that degree of competition, or with the amount of parental care the females gave. The answer turned out to be both. The more mating competition a species exhibited, as evidenced in the number of mates it was possible for a male to have, the higher the male mortality relative to that of females, even though females do not live as long overall. At the same time, the bias in mortality was higher when either sex gave a great deal of care to the young after hatching, lower when the chicks began to fend for themselves at an early age.

An even broader perspective on longevity in males and females comes from butterflies. Seemingly carefree as they flit from flower to flower, butterflies can exhibit at least as much grim determination as a bull elephant seal over mating, and to them the stakes are just as high as in their weightier relatives. Christer Wiklund, a Swedish biologist and Elvis fan given to quoting rock song lyrics in his scientific communications, and colleagues Karl Gotthard and Sören Nylin compared two species, the peacock and the comma. The peacock female generally mates with only one male in a very short breeding period. The comma butterfly has a rather more leisurely approach, with females mating repeatedly over a period of several weeks. This difference means that selection for males to live longer should be more intense in the extended mating species, because males that can hang on and live to mate another day will be more successful than those that put all their sperm in one basket, so to speak.

One of the many advantages of testing ideas in butterflies and other insects is that they can be brought into the laboratory in large numbers, unlike elephant seals, and the scientists did just that. It is also possible to determine how many times a female butterfly has mated by counting the number of sperm packages in her reproductive tract, a thoughtful convenience likewise not provided by mammals. As expected, the comma females mated more times over their lives than the peacocks did. How does this speak to the idea that sex differences in longevity reflect the evolutionary pressures on males and females? While comma male and female lifespans were virtually the same, male peacocks lived less than half the time that the females did.

Insects are also interesting here because they lack testosterone, one of the pathways to male immune inferiority. Nevertheless, male insects have the same pressure to compete as other kinds of animals, and while monogamy is rare in insects, they still vary in the degree to which males can monopolize mates and female mating is restricted in time.

We have studied two species of crickets in my laboratory, both native to Australia but with very different ways of life, something like the peacock and comma butterflies. The Polynesian field cricket, *Teleogryllus oceanicus,* enjoys a balmy tropical existence and breeds year-round, without a distinct peak in mating. Its close relative *T. commodus,* rather unimaginatively called the black field cricket, leads a harsher life in the chillier southern part of the continent as well as New Zealand, and it breeds during a period of several weeks in the fall. In both species, males call to attract females, who then go off to lay their eggs with no further input from the father. I wanted to know whether female crickets showed the same kind of immunological superiority seen in humans and other mammals, so we compared immune response in males and females of both species.

Somewhat to my surprise, when a sex difference appeared, it favored the males: Male black field crickets were better at resisting parasites, as evidenced by their superior ability to coat a tiny piece of fishing line in their abdomen with blood cells and other material. The tropical species showed no sex difference at all in this measure of immunity. I then began to wonder if the concentrated life history of the seasonally breeding black field cricket had made it particularly difficult for females to both resist disease and lay their eggs in a short period of time. Egg production is an extremely costly activity for a cricket; the eggs and associated tissues can comprise up to 20 percent of a female's body weight, compared with a mere 6 percent for males. Perhaps male competition wasn't the only thing shaping the way the sexes differed in disease susceptibility.

We then compared immune responses in a group of Polynesian field crickets that we had given a very restricted diet. Limited food should affect the crickets like limited time to produce eggs does, putting them under pressure to only do what's most important. Under those circumstances, the same pattern of better immunity in the males appeared. This suggests that indeed, the females are hard pressed to maintain their immune response and their reproductive ability at the same time, and when the season is short, evolution will always favor reproduction. While this explanation is speculative, it points away from identifying a single cause like testosterone in producing patterns of sex differences in disease susceptibility.

A GENDER GAP TRAP

The World Bank projected a closing of the gender gap in longevity by 2025, at least in developed nations, but I suspect this is wildly optimistic at the very least, and more likely utterly hopeless. The gap cannot close easily or quickly because it is the product of a complex framework shaped by evolution. This complexity is why no one will ever be able to point to a single proximate mechanism that always causes women to live longer, whether it is smoking, alcohol abuse, heart disease, infectious disease, or homicides. Regardless of any one immediate cause, the same evolutionary process that gave us men with beards has also, as Shettles pointed out a half century ago, made those men die earlier than women. This is not to say that we should give up and let males smoke, drink, or infect themselves to death. It is just that there is nothing "unnatural" about a sex difference in longevity, nothing that is due to a newfangled blip on the biological radar.

The evolutionary nature of the longevity difference also means that we should be able to predict where and when exceptions should occur. Take, for example, jacanas. Sometimes called lilytrotters because their long slender toes enable them to stand on lily pads, these shorebirds have taken conventional sex roles and turned them upside down. Females are half again as large as males, and fight vigorously among themselves for mates. Males do all the incubating of the eggs, and a female can defend up to four males in her territory, each of which has fertilized at least some, though not necessarily all, of the eggs she laid in his carefully guarded nest.

What about their disease susceptibility? No one knows. But I would wager that females, being the sex that has to live hard and possibly die young, would show the same pattern that is seen in males of most other species. It may well be possible to be male and not be the sicker sex. You just might have to live on a lily pond, eat weeds and bugs, and do all the childcare to get there.

CHAPTER 7

PARASITES AND PICKING THE PERFECT PARTNER

We were crouched in the shrubbery outside the men's bathroom in a park in Western Australia, binoculars at the ready. Periodically one of us would exclaim, "Wow, that was a really good one," or "Look at him now, he's really going at it!" Perhaps surprisingly, we were not accosted by the police, though we did earn some hard looks and hurried departures by the gentlemen availing themselves of the facility. While related to sex, however, our motives had nothing to do with their activities. We were watching bowerbirds.

Almost twenty different species of bowerbirds and their close relatives, the catbirds, live in Australia and New Guinea, and most practice unusually extravagant sexual habits. Males construct elaborate structures, ranging from foot-high U-shaped gateways to walled and roofed huts almost big enough to hold a bookshelf and a microwave and warrant a commission from a real estate agent. They use twigs, leaves, and other natural materials to make these bowers, which can take weeks to complete, and the entryway and interior are frequently decorated with leaves, shells, or berries, often in tones of a single color, like those terrifyingly minimalist apartments in upscale magazines where a room is bare of decor except for a lemon, a glass vase of daffodils, and a book with a yellow cover.

The bird my companions and I were watching was a Great Bowerbird, which arranges white objects—bones, shells, stones—

in an inviting avenue that leads up to a woven arch of twigs set under a bush or tree. Males acquire their decorations painstakingly, over many weeks, sometimes stealing them from neighboring bowerbirds, who in turn will make raids to steal them back. Our male had taken advantage of the nearby picnic area to avail himself of leftover chicken bones, which after time in the sun had weathered to a tasteful cream color. He wandered in and out of the archway, fussing with a protruding stick here, a stone there. But it wasn't until another bird showed up at the periphery of his estate that the action truly began.

To use a scientific term, the bowerbird went berserk. The visitor appeared to be a female, and the male immediately began a frenzied display that involved hopping back and forth through the arch, spreading a fan of lavender feathers on the back of his head, picking up and showing the choicest items in the bower, and producing sounds that we only believed were biologically rather than mechanically produced because we saw the male making them. Otherwise it sounded a bit like someone operating a very small chainsaw while rolling tin cans off of a corrugated metal roof. The problem was that the male couldn't simultaneously show the feathers, pick up objects, jump through the archway, and make the noises all at once. Instead, he shifted abruptly from one to the other, as if he had a series of frantic incoherent thoughts running through his mind, first "show her the bone, show the bone" followed moments later by "the feathers, she'll love the feathers" but then remembering "make the noise, make the noise, they always go for the noise" and finally "that's it, over to the bone again, or would the feathers be better?"

The object of all this passion watched in mild interest for a few minutes, head cocked to one side, and then hopped off on business of her own, with the male oblivious to her departure for some time. Nothing daunted, he resumed his pacing through the bower, as if mentally rehashing his performance and rehearsing for the next episode ("what was I thinking, stupid, stupid, *stupid,* next time I'll do the bone *after* the feathers, not before, it will be much more enticing after the feathers, or maybe the feathers both before *and* after, that's it . . .").

Females in this and other bowerbird species may visit several males before ultimately mating with one of them. After the consummation, the female leaves the bower for the rest of the season; it is not a nest or a shelter, merely a stage. She builds her nest, lays her eggs, and rears her chicks by herself somewhere quite removed from the male's bower. After the male's performance, his work is done. We know details of only a few species' lives, since not all are as oblivious to human activity as the Great Bowerbirds (who seem, at least based on personal experience, to have an odd predilection for setting up their bowers near restrooms and other public buildings), but in those that have been intensively studied, just a few males get most of the action. In Satin Bowerbirds, a species from eastern Australia favoring blue decorations, including plastic children's toys nicked from nearby yards, a single male once mated fifty-three times in a season, and up to twenty-five females may select the same male. A majority of the males put on their elaborate show in vain, and never attract a single mate.

Males of many animals put on similarly extravagant displays, though most do not make the same use of objects in the environment to assist their act. The male peacock, classic symbol of male allure among animals, lifts his cumbersome tail, spreads it wide and shimmies it gently, the eyespots on the feathers seeming to twinkle in the sunlight. Male harriers, a kind of North American hawk, climb skyward until they almost disappear from sight, only to plummet to the ground making a characteristic chipping call. Crocodiles attempt to attract females by roaring, often for hours at a stretch, and lizards do rapid push-ups that reveal the bright colors on their chests and bellies, which gives credence to the idea of the gym as a place where singles meet. Snakes and butterflies give off odors that attract the opposite sex, while nightingales, frogs, and crickets sing. Some types of spiders wave the tufts of hair on their legs in a stereotyped fashion, and if intrepid investigators shave off the fur, females ignore the stripped-down displays; stubble is sexy. Even tiny fruit flies strut and wiggle their wings, producing songs at frequencies undetectable to the human ear.

MORE THAN JUST A PRETTY TAIL

The evolution of all of these extraordinary traits, from the bower building to the songs to the odors, has puzzled biologists since before Charles Darwin. After all, producing and then dragging around a tail half again as long as you are is no small feat. It is costly because it can make the bearer clumsy, it takes energy to manufacture—food-deprived male animals almost always produce smaller or less elaborate ornaments—and because anything that conspicuous can also attract unintended recipients of the display, like predators. And of course males pay a price in longevity and disease susceptibility, too.

The solution that Darwin proposed is that such secondary sexual traits persist either because they are valuable to males as they battle over access to females (like the horns and antlers on many deer and sheep) or because females find them alluring. The first idea, that males compete for females, was immediately approved by Darwin's contemporaries, who probably found it agreed with their Victorian sensibilities. The second, that females exert an important force in evolution by preferring certain male characteristics above others, had a longer road to travel, but ultimately it too has been accepted, and females in many species have been shown to prefer flashier males. Those peacocks, for example, do not strut their stuff in vain; males with more eyespots on their tail feathers get more females than less-adorned ones, and if you take scissors to the tail and reduce the number of spots, females lose interest in the shorn males.

But there is still a problem, or more accurately, two problems. If the secondary sexual trait conveys information about the contributions a male will make to his offspring, so that a more colorful male is also better at feeding babies, then it makes sense for the female to scrutinize prospective mates for the best provider. Indeed, ceremonial feeding of mates is part of the courtship ritual in several species, and it is thought to indicate exactly this kind of doting parental intention. But what about the vast number of species, including the bowerbirds, in which females receive nothing from males other than the sperm to fertilize their eggs? What possible advantage could the female obtain by mating with a male who has built a more elaborate bower, or who has, in the case of the peacocks, a flashier tail, when neither she nor their offspring will ever see him again?

A popular answer is that the sexual ornaments indicate that the male has good genes to pass on to the female's young, so that a brightly colored male isn't just pretty, he is exceptionally fit, in the evolutionary sense of being able to survive in the environment. So males with elaborate secondary sexual characteristics are of higher quality, where quality means the ability to pass on genes to succeeding generations.

This is reasonable in principle. But genetic quality in a mate, like character in a politician, is a slippery concept. Do you mean being able to evade predators? Is finding food the most important attribute? What about resisting temperature extremes, or keeping from drying out if you are a frog living in ephemeral streams and ponds? Should the female look for an ornament that indicates all of these things, and if so, how could that work? Is it possible that a single set of tail feathers, a single evening's croaking from a frog, could convey so much information?

Even assuming a female could be so discerning, this leads to the second problem. Say she manages to select the male whose traits do indeed indicate his strength, intelligence, integrity, and ability to find worms. Her offspring will be similarly endowed, and both parents win, evolutionarily speaking. The male's genes will slowly begin to predominate in the population, and the females that choose him, and others like him, will likewise benefit. Eventually, however, the losers will simply be weeded out of the population. If males are only contributing sperm, and can mate with more than one female, as is virtually always the case for such species, then most of the females will pick the same "best" kind of mate. After some generations of this, all the women and men are strong and good-looking, and all the children, so to speak, are above average. Or more accurately, average isn't what it used to be. Everybody has gotten better and better, and any remaining differences among males in their secondary sexual characteristics are due to environmental differences as they were maturing; maybe one got caught in a storm during a crucial period for tail development and didn't get enough to eat, or maybe another simply got his back end caught in a fence. In any case, the genetic variability in fitness that provided the impetus for female choice is expected to disappear in time, leaving us with the same conundrum of why females bother to choose at all.

PARASITES TO THE RESCUE, AGAIN

As you might have guessed by now, an answer to both of these questions—what information do ornamental traits convey and why do females continue to choose—can be found in parasites and disease. In the early 1980s I was a graduate student at the University of Michigan, working under the late evolutionary biologist William D. Hamilton. Bill was in some ways the prototypical eccentric genius, with a shock of silver hair and a bemused manner. For the previous decade or so, Bill had been interested in, some might say obsessed by, the role of parasites in maintaining sexual reproduction, and he was one of the leading proponents of the Red Queen hypothesis. He also had no compunctions about coming up with unusual—some would say outlandish—ideas. When I arrived in Ann Arbor, he was just starting to think about how parasites might also explain some of the puzzles of sexual selection and mate choice, and my being an enthusiastic amateur bird-watcher fit into his plans nicely.

Bill thought that the best possible genes a female could look for in a male, and hence the best possible genes a male's ornaments should signal to prospective mates, are those that confer resistance to disease. Pressure from food shortages and floods may come and go, but parasites are always with us. If a female had to choose to learn one thing from the songs, smells, and dances, it should be about the health of her partner, and whether he is likely to pass on that health to any offspring he sires. The courtship antics of animals could therefore be viewed as a kind of peculiar medical exam, with a female examiner watching her suitor for signs of a lingering cough, a bad case of lice, or enough worms that growing a brilliant red crest becomes impossible. A link between ornaments and health seems more plausible and direct than that between ornaments and the ability to find shelter.

Furthermore, the interaction between host and parasite genes solves the problem of the loss of genetic variation that continued female choice seems to create. The same kind of arms race that evolved sex applies to mate choice too, with the genes for resistance to the pathogen on the part of the host, and for increased virulence on the part of the parasite, oscillating in stable cycles.

Analyzing such genetic cycles requires complicated mathematics. The point is that females do not need to know anything about any particular parasite, and indeed can ignore which genetic type is on top at the moment. All they need to do is choose the most resistant-appearing males, which they do using the aforementioned courtship medical exam. Certainly many diseases manifest themselves in the color and texture of the skin, the condition of the fur or feathers, and the vigor with which animals behave. Turkey breeders can look at the comb and wattles of their charges and diagnose tuberculosis if these fleshy parts are bluish, or fowl cholera if the wattles are swollen. Presumably a female turkey can be at least as discerning, since her genetic future, not just her economic one, is at stake.

Vigorous displays should also be good indicators of health and resistance to parasites, and indeed many sexual signals are very energy consuming. Male frogs, for example, rev up their metabolism several fold when they are calling at night. A bird that can't find food as well as his neighbor may or may not show it in his sway and swagger, but a high fever and gut full of worms are sure to dull his appeal.

Hamilton and I made two predictions about what animal sexual appearance and behavior should be like if our idea was correct. The first was simply that females from many if not most species should be able to use flashy ornaments and displays to determine which males have the fewest parasites, and should prefer to mate with them. We will come back to this one later. The second, less obvious prediction was that if parasites drive the evolution of showy colors, structures, and behavior, animal species that show an abundance of these traits—the most colorful, ornate, and dramatic ones—should therefore have been subject to the most parasites in their evolutionary history. If the function of a bright green tail or a blue eye or loud mating call is to show how parasite-free a male is, then the degree to which a species is plagued with parasites should mirror the degree of flamboyance its males must evolve. Therefore peacocks should have more parasites than wrens, and scarlet tanagers and blue jays should have more than sparrows.

This is where my interest in bird watching came in. Bill thought that if we could find lists of bird species and the abundance of their parasites, I could rank the males and females from each species on a scale from one to six, with one being really drab and six being extremely flashy. A male scarlet tanager or a northern oriole, then, would get a six, a Cape May warbler a five, and a house wren a two. In some birds, like many kinds of woodpeckers, males and females are quite similar, with only a small flourish of red on the head to demarcate the sexes, while in others, like the black-throated blue warbler, the sexes are so different that you could be forgiven for assuming they did not belong to the same species.

A few species posed special difficulties. Mockingbirds, for instance, look very subdued in the field guides, being gray and white, but in the flesh few birds can seem more ostentatious, as they flash their contrasting white wing patches and perch conspicuously on branches. Should I classify them as dull or give them points for personality, as it were? I went for the latter. What about the all-black or all-white birds, like crows or egrets? Is monochromatic boring, or, like an urban New Yorker, edgy and fashionable? I compromised, giving them more points than a mottled gray sparrow but fewer than the cardinals in their bright red plumage.

Next we needed to know how heavily parasitized each species

was. Here we were aided by a rather unusual avocation of bird watchers, or birders as they are usually called. Some birders are not satisfied with just seeing a species in the bush; they want it in the hand, literally, and they also want to find out at least some small piece of information about its life besides its identity. And so they start or join mist-netting programs. Mist nets are exactly what they sound like, nets with mesh so fine that it is barely visible against the sky, and they are set like volleyball or badminton nets about ten to twelve yards apart between poles that are two to three yards tall. Birds flying along the path of the nets cannot see them, and are caught in the mesh for the birder to disentangle, identify, perhaps weigh and measure, and sometimes—and this was the crucial point for our study—take a drop of blood by pricking a toe or wing vein. After they are measured, the birds are released unharmed, usually within minutes of their capture.

We were most interested in that drop of blood, which was usually placed on a glass microscope slide and examined for minuscule parasites. The blood parasites, while not usually fatal, can cause chronic debilitating illness in the birds they infect, and Bill and I suspected they would be good candidates for testing our idea, since a disease that simply killed its host or allowed it to recover completely would be less likely to leave its signature for prospective mates to read.

For reasons that I have never determined, collecting and preparing such blood smears, and examining them for parasites, was something of a cottage industry among biologists in the 1960s and 1970s. We found scores of papers that recounted their findings. Bill and I selected four sets of these surveys, using North American birds both because I was the most familiar with them and because the most information was available. Our final candidates were from Cape Cod, the District of Columbia, South Carolina, and Georgia, and Algonquin Park in Canada.

Finally, I had a friend who knew a great deal about birds rank the songs, much as I had the feather color. Song is a very important attribute of courtship in many birds, and species clearly vary in how much effort they put into their songs, with wrens, for example, being particularly musically gifted and many sparrows rather unremarkable.

Once all the data were collected, we put the scores for song and brightness along with the parasite information into a computerized program that would tell us whether or not there was a relationship between them. Much to our satisfaction, there was, and in the predicted direction: Brighter bird species had more parasites, as did those that sang more musically complex songs.

We were pleased to see that parasites were also more likely to plague proficient songsters, because that suggested that there wasn't some odd and spurious correlation between feather color and parasites. Bird blood parasites are carried by flying insects like mosquitoes that transmit the pathogens when they bite the host. Some skeptical colleagues had wondered whether the brighter species made easier targets, but it seemed unlikely that a gnat could tell the difference between a melodious song and an abrasive one.

Bill and I published our paper in *Science,* and it was greeted with a great deal of interest. But some of our colleagues thought the idea that parasites could influence the splendor of a peacock's tail seemed bizarre, while others felt that our method for scoring the birds was too subjective, and in one case a British biologist rescored the species, getting somewhat different results. Furthermore, even at the outset Bill and I were quite willing to accept that other factors in addition to parasites affect the brightness of a bird's plumage, and these other pressures will cloud our ability to see the relationship we predicted.

A bigger difficulty with our test arose when scientists began to be more aware of the pitfalls of comparing different species, whether they were looking at parasites, plumage, or intelligence. Bill and I claimed that the more species that showed the predicted correspondence between showiness and parasites, the more likely our hypothesis was correct. But if four kinds of tanagers have bright plumage and a lot of parasites, it is possible the tanagers all inherited their bright plumage and their susceptibility to disease from a common ancestor, back in their evolutionary history, and have all kept the same trait because of their close relationship.

New techniques in statistics and evolutionary biology have helped solve these problems, and even with corrections for this shared heritage, the relationship between brightness and parasites that Bill and I predicted still seems to hold. But the difficulties with testing the hypothesis by comparing species meant that biologists started looking harder at testing the within-species part of our idea: Females should be able to detect a resistant male using his ornaments, and would choose to mate with him over other types of males.

THE HEN'S DIAGNOSIS

I came to a fondness for chickens late in life. They had always seemed kind of embarrassing, like the geeky kid in elementary school with a perpetual runny nose who wanted to sit next to you at lunch. They lacked the cuteness factor of many mammals and the elegance of most other birds. I certainly never dreamed that they would provide me with some of the best support for our theory, and I definitely would not have thought that they would prove both endearing and fascinating in their own right. My history and that of the fowl began to entwine after I finished my doctorate.

I went to the University of New Mexico to work with David Ligon, an ornithologist with a lifelong fondness for poultry and game birds of all sorts, and Randy Thornhill, an evolutionary biologist who had mainly studied insects in his career. The two of them had decided to collaborate on a large-scale study of mating behavior, looking at all aspects of the topic, from what traits females prefer in a mate to whether males that are better fighters are also more likely to attract the opposite sex. They wanted to include the possible effects of parasites on mate choice, and hence asked me to participate. It all sounded great, until I found out that the object of the research was to be the red jungle fowl, a bird that is the ancestor of all of the breeds of domestic chickens we are familiar with today. Basically, I was being asked to study chickens.

A future in the barnyard is not exactly what I had had in mind when I got a Ph.D. (the phrase "and for this you went to college?" kept ringing in my ears), but postdoctoral positions were scarce, and both Randy and Dave are world-class scientists I was eager to work with, so I moved to Albuquerque. The jungle fowl we used came from the San Diego Zoo, where a flock had been introduced from Asia in 1942 and allowed to range freely over the extensive grounds.

Jungle fowl are not, strictly speaking, chickens; they have a number of attributes that set them apart from Henny Penny and other barnyard characters. Native to southeast Asia, they are quite a bit smaller than most domestic breeds, and the females look very different from the males. A jungle fowl hen is a study in subtlety. Her feathers are a tonal buff color and her legs are a modest gray-green, with none of the loud tasteless yellow found in domestic chickens. Her comb is small, and her tail is held low, straight out from the body.

The male jungle fowl, in contrast, is a thing of glory to behold. The ideal specimen has silky orange hackle feathers streaming down his neck, graceful sickle-shaped saddle feathers dangling over his legs, a crimson comb and wattles, and a brilliant orange-red eye that dares an intruder to come any closer. His tail feathers arc out behind him like iridescent scimitars, hanging nearly to the ground. Roosters are often portrayed as rather full of themselves, and fiction seems to be accurate in this regard, at least insofar as I could judge, but then a male in his prime has a great deal to be proud of. Chickens—at least in their primordial form—turn out to be beautiful, and they have complex personalities and lead lives full of drama. Shortly after arriving in New Mexico, I became a chicken convert.

My own rhapsodizing aside, what we wanted to know was what attributes female jungle fowl found most attractive in a rooster, and furthermore whether these were influenced by parasites. Here is where the genius of Dave and Randy's choice of subject shone. On the one hand, jungle fowl look and behave like the wild members of the pheasant family that they are, having not themselves been subject to the kind of agricultural breeding programs that have given us uniformly sized eggs and chicken breast meat of an exact shade of yellowish-white, not to mention hens that tolerate being in confined spaces and producing an egg per day, something that is completely unnatural for their wild ancestors. On the other hand, as my friends would never fail to remind me during episodes of the aforementioned rhapsodizing, the birds are nevertheless chickens at heart, or more precisely in their genes. This means that all of the information about chicken physiology, genetics, and rearing could be applied to our work, and we did not have to worry about inventing new methods for our experiments.

A case in point was the parasite we used, a roundworm that commonly infects the gut in domestic poultry and has been found in wild jungle fowl as well. It usually exerts its effects on the chicks, which can grow more slowly and show other abnormalities, but it is not immediately lethal, and hence like the blood parasites it seemed like a good candidate for investigating Bill's and my idea. Because of its frequent occurrence in chickens, poultry scientists could supply a great deal of information about the life cycle of the worm and how to handle it in experimental infections.

We infected a group of unsuspecting chicks by squirting saline containing worm eggs into their mouths. A control group got the salt water without the parasites, and then we let both sets of birds grow up and develop their secondary sexual characteristics. Chickens mature by about eight or nine months of age, after a gawky teenage stage that, while lacking problems with acne, is otherwise pretty much like human puberty. We then measured the size and color of the males' feathers and comb, the color of the eye, and overall body size.

Then we used the infected and uninfected males in mate choice tests, to see what the females noticed when they chose a mate. In nature, a hen will approach a male and solicit sex by crouching in front of him, and we wanted to allow this same behavior in our experiment. We didn't, however, want other males interfering with the process, and we also wanted to control which males the female was allowed to choose between. So we set up an enclosure with two compartments, and tethered a rooster by the leg inside each half, so that he could move around as if on a leash, but couldn't get out or see the rooster in the other compartment. The males adjusted surprisingly quickly to this restriction, and displayed and behaved pretty much as usual while tethered. We put one parasitized and one uninfected rooster into each of the compartments.

We then put a hen, which we hadn't infected with worms, into a pen outside the compartments, and let her watch the two males for half an hour. The males could see her, too, and they crowed and performed other courtship displays to attract the female's attention. After the time was up, we released the female and watched to see which male she approached. Most of the hens went straight to one or the other rooster and crouched in an invitation to mate. The roosters were virtually always happy to comply and we would score the female as having preferred that male. Occasionally there would be a mishap due to our set-up, when the hen would crouch just out of reach of the male; in these cases the hen would wait for a moment or two and then look back over her shoulder as if in puzzlement, waiting for the rooster to get on with it, while the hapless male struggled. We counted these attempts as a choice as well.

Armed with the measurements of the male's secondary sexual characteristics, we then looked first to see whether parasitized and unparasitized males differed in their appearance, and then if the females had chosen the uninfected rooster over the wormy one. The two groups were indeed different on both counts, though not exaggeratedly so—a male with worms could not be distinguished from an uninfected one all of the time, just as you'd expect to be the case in nature. In other words, our experiment hadn't created a class of incredibly sickly birds you'd have to be crazy to want to mate with. The females preferred the control birds by a margin of 2 to 1, and they discriminated based on the same characteristics that we'd shown in other tests using worm-free males to be important: comb size and color, eye color, and the color of the hackle or neck feathers.

At this point you might be thinking that after all of the convoluted theorizing and measuring, we'd really just come up with a common-sense result that any chicken farmer would have been able to predict from the outset, that wormy birds look worse and don't do as well at mating as healthy ones. "Elaborating the obvious" might come to mind, in fact.

But the chicken farmer would, I'd wager, be unlikely to suggest that worms should disproportionately affect secondary sexual traits, rather than traits that function outside of mate choice, like the length of a bird's legs or its body weight. If it's just parasites making birds wimpy and unattractive, they should be wimpy and unattractive all over. If, however, ornaments evolved because they show females which males are the healthy ones, then those are the traits that the parasites will target, and the traits that males will have a hard time faking unless they are truly worm-free. So I made another prediction: The infected and control groups should differ in the traits females use to decide which male to mate with, but they shouldn't in traits the females don't care about.

Here I simply compared the two groups of roosters using some characteristics that females had never paid any attention to in any of our tests, including length of the leg (which is a good indication of body size), beak size, and the length of the curved tail feathers. It turns out that you can't use these traits to discriminate between infected and uninfected roosters—the only traits that work are the same ones that females use to choose a mate. This kind of disproportionate effect is unlikely to be caused just because a chicken is sick, but it is exactly what you would expect if ornamental traits evolved to show females which mate is healthy.

These findings are far from the last word on our hypothesis. Many other tests in animals ranging from guppies to flies to mice have been done, and most of these, though by no means all, support the idea that females use secondary sexual traits as a way to get parasite-free mates. Even the bowerbirds seem to show some effect of parasites on their extravagant displays.

One of the best studies was by Danish scientist Anders Pape Møller, who used barn swallows, the graceful birds with long forked tails. He was able to show not just that females preferred to mate with longer-tailed males, and that a bloodsucking mite made the males' tails grow in shorter, but also that the resistance to the effects of the mites is passed from a healthy long-tailed male to his offspring. This genetic component to the whole process is an essential part of our idea, but it is often difficult to demonstrate. In the swallows, both parents help feed and care for the chicks. What if, for example, long-tailed and parasite-free males simply maintained the nest so that it was likewise free from parasites, but did not pass on the genes for resistance? That would make it look like our hypothesis was supported, but according to our idea, the females have to be choosing resistant males because they will pass those genes on to their offspring, even if those offspring never see their father. Møller switched swallow chicks from one nest to another, so that some of the babies from parasite-free fathers were raised by parents that were infected themselves. He found that mite loads of the offspring were more similar to their genetic fathers, rather than the foster parent, and that long-tailed fathers had parasite-free chicks. This means that at least in this species, females can indeed get genes conferring resistance to their babies by mating with an extravagantly ornamented, and healthy, male.

As with the comparisons of parasites among different species, and for that matter as with the role of parasites in explaining the evolution of sexual reproduction itself, parasites are not the only answer to the question of what secondary sexual ornaments are all about. But they are another reminder of how much a part of us disease has become, with its imprint evident in the brightness of a gaze, the curl of a tail, or the notes of a song.

LICE, HAIR, AND GETTING DATES

What about humans? While not as impressively ornamented as your average peacock, men and women still look different from each other, and it is reasonable to ask whether our own secondary sexual characteristics, like beards and breasts, could indicate our parasite resistance to a prospective mate. When I talked about the jungle fowl work to public groups like the Audubon Society, the audience invariably wanted to know what our results meant for their own love lives. (They also always wanted to tell me chicken anecdotes, including the story from one eighty-five-year-old woman with inch-thick glasses about her two pet chickens from girlhood, named Brother and Sister. Sister, she said, would wait outside the gate for her return from school, which I was quite willing to believe, and also was housebroken, which I absolutely did not. But that is another story.)

It certainly makes sense that people want a mate who is healthy, just like it makes sense that they want one who will be able to take care of their children. But applying Hamilton's and my idea to human mate choice is dicey. For one thing, people are unlike the bowerbirds, jungle fowl, and guppies, because males stick around after mating, at least some of the time, to provide for the offspring. This means that the puzzle of why females would bother to choose a particular male is less of an issue in our own species than it is in those where males provide only sperm to the female. A woman doesn't have to frantically scan the beard development of a man on a single date to try and figure out if he is resistant to malaria and whooping cough; she can get numerous clues about his health, along with his likelihood of earning a good living and laughing at her jokes, during a more prolonged courtship. And given our at least ostensibly monogamous society, men can do the same thing by examining women. In either case, however, good genes are only part of the picture.

That part, of course, could still be important, and Bobbi Low, a biologist at the University of Michigan, looked at human societies around the world to see whether parasites were more prevalent in cultures where polygyny, the practice of a single man having many wives or concubines, was more common. She reasoned that this was similar to our prediction about heavier parasite burdens leading to higher degrees of sexual selection and hence showier plumage in the birds. Indeed, societies where polygyny is common are more plagued by diseases such as malaria, sleeping sickness, and various worms. But this association could arise for reasons other than our theory, including the rather mundane one that where parasites are common, wealthy men might be able both to afford more wives and keep disease at bay. No human society has the equivalent of a bowerbird mating system, where females are completely free to choose and males display and hope for the best, though some bars and clubs may come close.

One link between the bowerbirds and us might prove promising, however. Two British scientists, Mark Pagel and Walter Bodmer, suggested that mate choice on the part of both sexes to avoid parasites has caused our most obvious difference from our primate relations: hairlessness. In an article succinctly titled "A naked ape would have fewer parasites," they reasoned that less body hair gives lice, fleas, and ticks fewer places to hide, and so relatively less hirsute individuals could better advertise their vermin-free state, and hence would be more desirable as mates. Once fire and clothes had been invented, it would have been possible to compensate for the loss of protection offered by fur, and so hair was kept only on the head, pubic region, and armpits because it afforded an opportunity to concentrate odors that communicate other information about sex. Lice and other external parasites are more than a nuisance and bane of the parents of schoolchildren; they can carry serious diseases, such as typhus, and in large numbers can drain enough blood from the host to cause anemia. Whether this drove our evolution as a naked ape is unknown, though Pagel and Bodmer's idea is in my opinion at

whether or not a male has parasites at any one moment has the potential to confuse exposure to disease, which can depend on sheer chance, with resistance to it, which is the quality the female is seeking. To really gauge whether a male with bright colors and a loud song will deliver healthy genes for his offspring, females need to know if those traits reveal a robust immunity.

THE TRUTH ABOUT TESTOSTERONE POISONING

How might the females do that? To make a connection between immunity and looking good, it is necessary to travel above the Arctic Circle, nearly to the top of the globe, where a scientist named Ivar Folstad works at the University of Tromsø in Norway, the northernmost university in the world, where it is dark twenty-four hours a day for several months during the winter. In 1992 he and his colleague Andrew Karter published a paper explaining just why traits like bright colors are expected to signal the strength of the immune system. The link occurs through testosterone, the hormone with that dark underbelly of immune suppression.

In addition to its effects on death rates, testosterone acts as a double-edged sword in the mating process. In an ideal world, a female choosing a male rooster because he has a big comb means not only that her offspring have big combs, but that they inherit the disease resistance that his big comb indicates. But what happens if the genes for the big comb don't always co-occur with the genes for disease resistance? Then a male could cheat—produce the big comb and gain the attention of females without having the good genes to back it up. A female could no longer rely on the comb to tell her about male quality.

Here is where testosterone comes in. If testosterone is necessary for the production of the secondary sexual characteristics that males need to get mates, but it also increases their vulnerability to disease, only those males of particularly high quality would be able to maintain showy ornaments despite the onslaught on their immune systems. Females therefore have a foolproof way of detecting the real studs, because a male who cheats by producing a long tail or big comb, but is not also of high quality, will be unable to pay the price of compromised immunity and will succumb to disease. The link between testosterone and immunity keeps males honest.

To test Folstad and Karter's hypothesis, scientists wanted to look at a variety of animals to see if testosterone, immunity, and ornaments were connected. Measuring ornaments is a snap, or at least something behavioral biologists are used to doing. Assessing testosterone levels, or manipulating them experimentally, is a bit more difficult, but almost as feasible in animals as it is in athletes. We can elevate testosterone levels in birds, for example, by placing thin plastic tubes filled with a powdered form of the hormone just underneath the skin, where the testosterone slowly leaches out into the bloodstream.

Measuring immunity is a different story, particularly in wild animals living under natural conditions, far away from the sterile instruments and complicated machinery that is de rigueur for an immunology laboratory. Furthermore, the vertebrate immune system is complex, to say the least; which elements does one measure to get an overall idea of disease resistance?

IMMUNOLOGY FOR DUMMIES, OR AT LEAST FOR ECOLOGISTS

Before deciding what to measure, ecologists and behavioral biologists found themselves trying to take a crash course in immunology. At least from my perspective, this has proved to be a daunting task. Immunology is filled with more jargon and complicated explanations relying on historical accidents than any other specialty in biology. On the other hand, you have to love a field that uses the word "idiotype" in its lexicon, even if it turns out, somewhat to my disappointment, that an idiotype is not a type of idiot. (It is a kind of immune molecule defined by the structure it uses to recognize foreign substances.)

A few salient principles, however, are reasonably straightforward. All immune systems do the same thing: distinguish self from nonself. Nonself can mean a bacterial infection, a worm in the gut, a cell laden with flu virus, or a transplanted organ. Transplant recipients have to take powerful drugs to suppress their natural tendency to reject the organ, which also renders them more vulnerable to infectious diseases. Exactly how the body responds once a substance is identified as nonself varies, but in most vertebrates the immune system can be roughly divided into two components, cell-mediated and humoral, which is also called antibody-mediated.

White blood cells are subdivided into several types, and cell-mediated immunity relies on T cells, produced in the bone marrow but matured in the thymus, a small organ located just below the neck. Cell-mediated responses are quite general, a kind of one-size-fits-all method of helping with wound healing, consuming damaged cells and debris, and orchestrating the inflammation that often accompanies infection. That inflammation is helpful because it increases blood flow to affected areas, and blood is how most disease-fighting substances get around. T cells are particularly important in fending off tuberculosis and several other diseases, and they are also targeted by HIV, which is why people with the virus watch their T cell count; a low number suggests that the virus is gaining the upper hand, while a higher count indicates that for the moment the infection is kept at bay.

Cell-mediated immunity can be measured in wild animals in a few ways. One of the most commonly employed methods in birds takes advantage of the role of T cells in inflammation. A tiny amount of a foreign protein is injected just underneath the skin of the wing, in the "web" between the two main bones. The protein is not a disease-causing organism, just something the bird's immune system will recognize as nonself and generate an inflammatory response. Swelling at the site of the injection reflects the overall readiness of the T cells, so birds with relatively larger swellings have better cell-mediated immunity.

Humoral immunity (named originally after the fluids or "humors" of the body—remember, I said this stuff was arcane) is more specific, involving the production of antibodies. Antibodies are molecules produced after a particular foreign substance is introduced into the body, and they are uniquely suited to recognize and lock onto that substance and eventually make it harmless. Once antibodies to a particular disease have been produced, they exist in a reserve state for an extended period, often for life, so that if the disease agent enters the body again, it is swiftly dispatched before it can multiply to dangerous proportions. The white blood cells most important in antibody-mediated immunity are the B cells, which in mammals are produced in the bone marrow. Vaccination works via humoral immunity; an altered version of a disease-causing organism is introduced to stimulate antibody production without actually causing illness, and then when the "real" pathogen arrives, the means for defeating it is already in place and the host doesn't even know it was ever at risk.

Gauging humoral immunity is trickier than doing so for cell-mediated immunity, partly because there are more components to measure and because knowing that an animal can swiftly produce antibodies to one pathogen does not necessarily mean that it can do the same thing for another. Biologists do some of the same things to wild animals that a general practitioner will do in an annual physical, such as monitoring white blood cell counts or looking at the types of white cells. They may also use a modified vaccination procedure, in which a foreign substance is injected into an animal and the antibodies specific to that substance measured. It is sometimes difficult to draw conclusions because the immune system bears witness to past infections, and an animal with a vigorous response to one protein, or an elevated count of one kind of white blood cell, could be indicating either a vigilant immunity or the weary track record of an infection-ridden past.

The two arms of the immune system interact in many ways, and a bewildering array of molecules and messengers are needed for the system to function. Other organs, such as the spleen, are also involved, and different groups of animals have somewhat different structures for dealing with disease. It is undeniable, though, that mounting this response takes energy, which for many animals is in short supply, lending further support to the idea that males capable of handling these tasks while growing a fine set of antlers are the high-quality mates females do well to choose.

OF SONGS, SPLEENS, AND COMBS

Armed with this information about the immune system, biologists began to try to see whether females could determine the quality of a male's immune system by looking at his secondary sexual characteristics or watching him court. Birds are favorite subjects, both because they have obvious ornamental traits and because their hormone levels and immune systems are relatively easy to monitor. One of the best examples comes from research using the European starling, a songbird native to Europe, Asia, and North Africa but introduced to North America in 1890 by a misguided soul who thought it would be nice if all of the birds mentioned in Shakespeare could be seen in the New World. Although only a few of the one hundred starlings released in New York's Central Park survived, over the next century they managed to reproduce a million-fold, and they now represent a threat to many of the native North American bird species. This rather nefarious side to their nature aside, starlings are gregarious, raffish birds that easily become tame in aviaries. The males have loud and variable songs, often mimicking other birds, and if tutored can even learn to talk, although their skills do not equal those of parrots or mynahs. Bird song is controlled by a few distinct portions of the brain, and these become larger under the influence of testosterone, supporting the notion that songs are ornaments just like tail feathers.

Deborah Duffy and Greg Ball from Indiana University decided to study starling song in the context of immunity. They recorded the songs of sixteen male starlings that were placed in an aviary with a reproductively receptive female starling, and then noted both the number of bouts of continuous song and their length, making the reasonable assumption that more vigorous and enthusiastic songs are the equivalent of a more ornate secondary sexual characteristic. They then tested both cell-mediated and humoral immunity in the birds, and found that the number of song bouts was higher in birds with better cell-mediated immunity and the length of the bouts greater in birds with better humoral immunity.

In another study of the connection between ornaments and the aptitude test of immunity, two biologists from the United Kingdom, François Mougeot and Steve Redpath, took ingenious advantage of the red grouse hunting that takes place each fall in Scotland. They examined the shot birds to determine intestinal parasites, comb size, and the weight of the body and spleen (birds with smaller spleens have had less cause to defend themselves against parasites). They also captured twenty-seven males, gave half of them a drug to reduce their intestinal worms, and performed the same measures on them, although they measured cell-mediated immunity rather than killing the birds to examine their spleens. Males with bigger combs were in better condition, had lighter spleens, and better immunity, suggesting that females choosing those males got a high-quality mate potentially able to resist disease and pass that resistance to their children. Mougeot and Redpath and several other colleagues also gave testosterone implants to some of the red grouse males while they went about their business in the field. They found that the birds with these artificially higher hormone levels grew larger combs, but also had more one-celled parasites, consistent with the double bind testosterone can produce for males.

HOW THE RICH GET RICHER

Meanwhile, we were doing work of our own on immunity in the jungle fowl. I wanted to understand how immunity fit into the social lives of the birds, and because we kept them under semi-natural conditions outdoors, we could watch their behavior. In particular, I wanted to see what life was like for the head guy, the rooster in charge, the alpha chicken. The idea of a pecking order or dominance hierarchy arose when a Norwegian biologist named Thorleif Schjelderup-Ebbe watched flocks of barnyard chickens in the 1920s; he came up with the idea of assigning Greek letters to the animals based on their place in the hierarchy, so whoever had the highest status was alpha. (He also based his studies on hens, not roosters, which means that the original phrase should be alpha female, not male.)

In any event, chickens—and jungle fowl—form hierarchies within each sex, and in their native habitat in southeast Asia, the jungle fowl live in groups that contain a dominant male, a few subordinate males, and several females with their chicks. The dominant male mates with the females most of the time, and defends his position as well as his flock with zeal; cockfighting is a widespread sport, and the chicken was originally domesticated not for eggs or meat but for this fierce aggression. Females do choose males, as I described, and the long comb that is the badge of a desirable mate is also key in maintaining dominance over other males. Roosters with large combs tend to win fights, almost regardless of body size, perhaps because the high level of testosterone necessary for the ornament also fuels the male's pugilistic tendencies.

My question was how these roosters juggled the demands of social competition with the need to defend themselves against disease, given the conflicting effects of testosterone. So we took roosters and kept them in individual large outdoor cages, each with a single hen. The males could hear each other crow, which they do often and vigorously (those who think dawn is the only time roosters crow have never been acquainted with chickens in real life), so they knew, so to speak, that other members of their species were around. This treatment has the effect of equalizing each male's sense of his own superiority, because roosters use the frequency with which other males challenge them while they have a hen nearby as a gauge of their status. If no one messes with you on a date, the chicken reasoning seems to go, you must be pretty hot stuff. They seemed to not notice the cage wire, apparently attributing their solitude to recognition of their supremacy. No one ever claimed chickens were intellectual giants. This meant that we could start with a group of birds who all had the same level of self-esteem.

We measured immunity and comb size in the males while their experimentally induced confidence was high, and then played what amounted to a mean trick on them: We put them into a new group of two males and three females, and waited for the males to establish a new hierarchy. Now the roosters had to fight to maintain status against a real opponent, and in the barnyard, someone always wins and someone loses. The ranks were usually established within a few days of pecking and chasing, with no serious injury, and after the metaphorical and literal dust cleared, we measured immunity and combs again.

The first thing we found was that males with bigger combs before the flocks were formed were more likely to end up dominant, consistent with our previous observations that males with big combs win fights. More interesting was the effect, not of comb size on status, but vice versa: The combs of males who ended up dominant grew, while those of the loser males actually shrank a few millimeters, not a huge difference, but noticeable. Intimidation can change your looks, even if you are a chicken.

What effect does it have on your immune system? The birds that became dominant in the flocks turned out to have had better immunity than the eventually subordinate birds even before the two of them were put together, so they started out ahead. Surprisingly, after the presumably stressful fighting was over, the dominant males were even better than the subordinates, a kind of rich-getting-richer phenomenon. And among the dominant roosters, longer combs meant better cell-mediated immunity, whereas the opposite was true in the losers, where longer combs were associated with worse cell-mediated immunity. It is possible to have it all, if you don't mind "all" being defined as a big red fleshy protuberance on your head and a lot of white blood cells. But isn't that against the rules?

I think these results can be explained by what evolutionary biologists call the car-house paradox. If you assume that people's funds are finite, which is reasonable, then any money spent on one thing is unavailable to spend on anything else, as many credit card holders are chagrined to discover at the end of the month. You would therefore expect a tradeoff between how much is spent on, say, a car, and how much on a house, with a negative relationship between the two, so that people who spend a lot on a car have little to put into their houses and vice versa. But of course the opposite is true; with few exceptions, Mercedes drivers do not live in hovels, and people with ten-year-old Hyundais tend to have modest houses. The reason for the apparent contradiction is that people start out with vastly different amounts of money, so that even though on an absolute scale a millionaire has fewer dollars available for a mansion once that Hummer is in the garage, the tradeoff is imperceptible compared with someone who does not even earn the price of the Hummer in a year.

The poultry equivalent of being a millionaire is being of high enough quality that spending the energy on your comb doesn't mean you are bankrupt when it is time to maintain your immune system. You can have them both, and a female who notices that is in a good position to get the best father for her chicks. By keeping their combs big while another male is around, the dominant birds were in effect advertising their ability to pay the cost of having an ornament that incites combat. The physiological cost of producing that few millimeters of tissue is probably negligible; the real cost is the social one, of broadcasting to all comers that the rooster is a contender. Doing so while keeping up the fight against pathogens is daunting indeed.

WHY IS TESTOSTERONE LIKE DOUGHNUTS?

Discussions of the dark side of testosterone almost always make men—though rarely women—ask how such a cruel bind could have evolved. It seems to be the same way people feel when they find out the nutritional profile of junk food: How can something that tastes so good be so bad for you? Why has evolution played what seems like a mean-spirited joke on all male vertebrates?

Part of the answer to this question is the same as to any other question about the imperfections of life; evolution seldom results in the best of all possible worlds. Remember Nesse's point that natural selection maximizes reproduction, not health, leading to bodies filled with compromises. It would be nice if testosterone not only gave males beards and sperm but made them more resistant to disease, just like it would be nice if walking on two legs made childbirth easier instead of more difficult. But that is the way the Darwinian cookie crumbles. There may simply be an unavoidable link between great antlers and lousy T cells.

A more subtle reason for the dilemma may lie in the blindness of natural selection to the future that I already discussed. Imagine a gene that makes males superb fighters and devastatingly attractive to females. Males with that gene will sire more offspring than other males, and therefore such males will become more prevalent in the population. But say that gene also makes males more vulnerable to disease, so that they die at an earlier age than the other males. As long as they still outreproduce their less sexy brothers, the gene will persist, because it acts while more members of the population are still alive, and by the time the deleterious effects roll around, many individuals would have died of other causes. It is nearly always better to get the benefit now rather than delay gratification, from an evolutionary standpoint at least, since a gene that only helps you reproduce better at a late age stands a good chance of never getting to exert its effects. So the price of testosterone may be worth paying from an evolutionary perspective, even if it seems like a raw deal in the here and now.

Finally, remember that even if males suffer from the effects of testosterone, females may gain from it. If only high-quality males can maintain their ornaments, testosterone gives females a cheat-proof signal for finding a good mate. It is possible that this benefit to females is sufficient to override the detriment to males.

NEVER MIND THE BIRDS; WHAT ABOUT THE BEES?

All of this concern about testosterone and its effects is very narrow-minded, not because it focuses too much attention on males, but because only a very small minority of animals have testosterone to begin with. The hormone is found only in vertebrates, and therefore the vast majority of animals on the planet, both male and female, get along fine without it. Testosterone poisoning, real or imagined, is not a problem for insects, spiders, crabs, worms, snails, and the other millions of species without backbones.

Nevertheless, these animals still have the conflicting demands of defense against disease and reproduction, and furthermore females in these animals would still benefit if they could determine which males had genes for resistance that could be passed on to offspring. Invertebrates are plagued by a wide variety of diseases. Although it may be hard at first to imagine what a sick butterfly looks like, myriad fungi, bacteria, viruses, and worms invade the bodies of animals without backbones just as they do vertebrates. A sick butterfly may not have the sniffles, but it can become lethargic, its organs can deteriorate, and it can suffer in many ways both obvious and invisible to the observer. Invertebrates also have the virtue of possessing much simpler immune systems than vertebrates, making them a great deal easier to study. What is more, your mate choice carries equally drastic consequences whether you are a cricket or a fly or a bird, or for that matter a person, at least in some respects.

Some of the best examples of this influence come from bumblebees, those furry relatives of the honeybee that motor determinedly from flower to flower and live in small colonies of perhaps a few dozen individuals, far less than the honeybee metropolis of tens of thousands. A colony has a single queen, and each queen has a single mating episode in her life, after which she retires and produces eggs for the other members, her daughters, to rear. A single male bumblebee can probably produce enough sperm to fertilize all the eggs she will lay in her lifetime, and bees and ants are uncommonly good at storing sperm for long periods, up to several years in the case of some species. Nevertheless, the bumblebee queen will sometimes mate with a number of different males before she settles down to a life of celibacy. Other colony-living insects like ants and wasps show similar patterns.

What is the advantage of this multiple mating? In the bumblebees at least, it seems to serve as a bet-hedging device against disease. Members of colonies where the queen mated with more than one male are more resistant to a common intestinal parasite than colonies in which the queen was monogamous, and Swiss scientists Boris Baer and Paul Schmid-Hempel demonstrated that the resistance came from the father's genes by artificially inseminating queens with sperm from one male, two unrelated males, four brothers, or four unrelated males. Doing this successfully without being stung is quite a feat, as you might imagine, but like anything else one gets better at it with practice. It turned out that worker bees from different fathers differed dramatically in their susceptibility, more so the fewer genes the fathers shared, suggesting that queens mating with multiple males increase the likelihood that at least some of their offspring will do well in succeeding generations.

Studies of other insects show that male immunity is linked to sounds, smells, and visual ornaments. Cricket song can contain information about a male's immune system. Male beetles with better immune systems produced odors that were more attractive to females, and male damselflies showed darker spots on their wings, an important component of territory maintenance, when they had better immunity.

LOVE HURTS

The champions of a connection between immunity and mating, however, are the species where the phrase "wounds of love" takes on a new meaning. First, let us consider bedbugs. These insects used to be unsavory indications of less than hygienic bedding and were seldom seen in modern society, much less talked about. They have made a perplexing recent comeback in urban areas, including New York City. Bedbugs are insects that make their living by sucking blood from their slumbering hosts, and while they prefer to use humans, other mammals or even birds will do in a pinch. They travel well, snuggled into suitcases, clothing, or cracks in furniture, can go for many days without feeding, and have a virtually cosmopolitan distribution.

Their sex lives, however, are what make them the focus of our attention here. When bedbugs mate, the female does not use the natural opening into her reproductive system to receive the sperm. Instead, the male punctures her body with a needle-like structure called a paramere, in a process rather aptly named traumatic insemination. The sperm then swim through the female's body cavity to fertilize her eggs, leaving a wound in their wake. Why this group of insects has evolved such an extraordinary method of reproduction is not clear. From a male's perspective, ensuring that his sperm not only enter the body of the female but make it to fertilization is paramount. From the female's point of view, however, it is beneficial to control the paternity of her offspring, and sometimes shun less-favored males as fathers even after she has accepted their sperm. The bedbug system may have evolved as a mechanism for males to circumvent such female control over the fate of sperm.

Regardless of the origin of the behavior, the effects of the wound incurred during copulation are dramatic: Mated females do not live as long as virgins, and females that mated five times produced nearly 25 percent fewer eggs than those mating only once, an evolutionary trauma indeed. Furthermore, when her body cavity is pierced in this manner, the female is left vulnerable to all kinds of potentially nasty bacteria and other microorganisms that can enter through the injury. Her vulnerability is only increased by the unsanitary conditions in which bedbug coupling takes place; when not feeding, the insects spend their time crammed into the crevices of the furniture and floors of the bedroom, and these areas are, as Klaus Reinhardt and his colleagues at the University of Sheffield in England observed, "replete with exuvia [previously shed skins of growing bugs], dead bugs, and the digested remnants of their blood meals." To say the least, this is not a propitious site to expose an open wound.

Female bedbugs, however, are not entirely defenseless. Males always pierce the female's body in the same place, on the abdomen, and when they do so the paramere has to pass through a special anatomical structure unique to female bedbugs. This structure contains blood cells that contact the sperm before it proceeds to the female's ovaries. Reinhardt wondered if the structure, called a spermalege, might function to ameliorate the transfer of germs into the body of the female. He and his colleagues used a glass needle to mimic the male's paramere, and they stabbed groups of female bedbugs either through the spermalege or through the abdomen nearby. In half the piercings, the needle was sterile, and in the other half it was first dipped in a solution containing bacteria. The scientists predicted that if the spermalege helps the female avoid infection, the sterile needles should be less harmful than the bacteria-laden ones, but the contaminated needle should have more damaging effects when it is inserted into the body directly, bypassing the special organ. Indeed, the females survived longer, and laid more eggs, when they received a simulated germ-laden mating through the spermalege than when the needle and its contents went into the body cavity. The spermalege is not perfect; uncontaminated punctures resulted in still higher survival and reproductive rates. But the evolution of an entirely new organ of the body that ameliorates the effects of pathogens acquired via mating suggests that sex and disease are intimate partners indeed.

Another kind of insect can provide the assurance that the wounds of love are equally distributed between the sexes. In a group of small crickets called ground crickets, mating is accompanied by a rather unusual activity on the part of the female. As most crickets do, she climbs onto the male's back and positions herself to receive the packet of sperm that he extrudes into her reproductive opening, a painstaking process that requires he thread the end of the packet into her genitalia so that the sperm can pass into her body. Unlike most crickets, however, the female ground cricket chews on a spur on the male's hind leg while this is occurring. Careful work by biologist Ken Fedorka showed that she is actually sucking the male's blood during this time, snapping off the tip of the spur as if it were the end of a drinking straw. The male can lose an appreciable amount of weight during these vampiric events. The bloodletting is not in vain, however; Fedorka found that the bloodthirsty females benefit from their consumption by increasing the number of eggs they lay the longer they sip.

My interest in this series of events was not in the gore, but in what the spur-chewing meant for the immunity of the male. Any time the external skeleton is pierced, an immune response is triggered, probably because in most cases a puncture signals the invasion of a parasite or bacteria. But the ground crickets are making love, not war, so the signal that an invasion has occurred might trigger an entirely unnecessary and costly response. Can the males get around this unwitting juxtaposition of sex, which is good, and signs of sickness, which are bad?

I have been working on this problem in my laboratory (at times I feel like the Anne Rice of crickets), and early results suggest that the males may be stuck. The more times a male mates, the worse his immune system becomes. And the decline has a price; at a given age, males are more likely to die, even in the laboratory under the most coddled conditions, the more times they have mated. Presumably the loss is worthwhile, since not reproducing is a fate worse than death, evolutionarily speaking, but it is not trifling. Males may be able to make a preemptive strike by shoring up their blood cell numbers and antibacterial compounds just as they mature, but that too has its price. Disease, and defense against it, has fashioned the sounds, sights, and scents of sex, even without testosterone to urge it on.

CHAPTER 8

WHEN HOW YOU FEEL IS HOW YOU LOOK

"I'm tired of all this nonsense about beauty being only skin-deep. That's deep enough. What do you want—an adorable pancreas?"

—*Jean Kerr,* The Snake Has All the Lines

Jean Kerr has a point. The superficiality of beauty is itself over-rated, and looking gorgeous on the outside ought to be good enough for all of us, thank you. At the same time, while no one needs an adorable pancreas, a properly functioning one is high on the list of desirable commodities. And having beautiful children who also have disease-free organs is every parent's fondest wish. In one of those teen-magazine hypothetical dilemmas, which would you choose: blooming health or stunning beauty, assuming you could have only one? It's not so easy to decide; life is undeniably easier for good-looking people.

But what if you didn't have to choose? What if being beautiful meant you were healthy? What if it also meant that you were eating the right foods, which suggests that you know where to forage and how to elbow competitors out of the way? That would then give the characteristics we use to define beauty—legs, lips, feathers, or musculature—some real significance, and make a preference for red skin patches rather than yellow, or curly hair over straight, far from arbitrary. Or is the rooster's comb really a red herring?

———

WHY REDDER IS BETTER FOR RADICALS

Millions are spent every year on the magic ingredients that promise to make or keep us healthy—fiber in oatmeal, omega-3 oils in salmon, or lycopene from tomatoes, not to mention the various herbs, teas, and tinctures said to prevent colds or build bones. Magazines are full of admonitions to eat properly for wrinkle-free glowing skin, shiny hair, or strong fingernails. Even animals are getting into the act; poultry farmers, dog breeders, and zookeepers all know that what an animal is fed influences how it looks both inside and out. Give a chicken lots of vitamin A in its feed, and the skin and meat will look appetizingly golden. Ads for pet food promise shinier coats and brighter eyes, which presumably make the animal more appealing to show judges and owners, if not prospective mates.

A large proportion of the traits animals use in mate choice are also diet-dependent, and the link between beauty, diet, and disease is one of the most exciting things to emerge in animal behavior over the last several years. Antioxidants in vegetables or vitamin supplements are the latest health craze, but people may not realize that they are also essential for the production of sexual ornaments in many animals. Carotenoids are the pigments responsible for the pink of a flamingo and the eponymous hues of salmon, vermilion flycatchers, and scarlet tanagers. They also make duck bills yellow and guppy scales orange. Plants and fungi produce more than six hundred different kinds of carotenoids, but no animal, including humans, can make them; everyone has to obtain them from food. We get carotenoids from a wide variety of fruits and vegetables, including carrots, tomatoes, and plums, as well as from products from animals that have themselves eaten carotenoid-rich foods, like eggs from chickens. The sources do not even have to be red or yellow themselves;

green leafy vegetables have high amounts of certain types of carotenoids.

In addition to making colorful skin, feathers, and scales, carotenoids are important parts of the immune system. They mop up damaging atoms or molecules called free radicals. We generate free radicals when we digest food, especially cooked food, and when we take in pollutants, tobacco smoke, and pesticides. Through a process called oxidation, free radicals injure cells, and more free radicals in the body are associated with nerve cell damage and other diseases, as well as accelerated aging.

Carotenoids are antioxidants, and they can detoxify the free radicals by stabilizing them with an extra electron. They also seem to stimulate other components of the immune system to function more efficiently. Diets high in antioxidants are touted as cancer preventatives, though controversy rages about whether we benefit from ingesting the compounds in supplements, versus in their original plant or animal sources. In addition, some diseases directly interfere with the utilization of carotenoids within the body.

Given these findings, it is logical to suggest that animals with sexual ornaments that require carotenoids for their showy colors are therefore signaling both their foraging ability and their health with those ornaments. If you can make a red feather only by eating the right foods, then only those capable of finding enough of those foods in their environment would be able to boast bright plumage. The presence of the colors in turn would imply healthiness, thanks to the high levels of carotenoids. John Endler, an evolutionary biologist at the University of Exeter in England, first proposed this connection for the guppies he studied in streams and rivers, where bright orange patches on males are alluring to female guppies. George Lozano, a Canadian biologist, then had the idea that if females looking at males focused specifically on traits that require carotenoids for their fullest development, like red feathers or skin, they would get a mate that also had immune-enhancing antioxidants.

This all made sense in the abstract. Starting in the 1980s, biologists began to scrutinize animals to see whether this idea worked in the real world.

IF IT'S GOOD ENOUGH FOR VULTURES...

Vultures are unlikely to win anyone's beauty contest. Most of them have unrelentingly drab feathers, sometimes livened with naked patches of yellow or red skin around the eyes and bill. Even the impressive California condor, its wingspan nine and a half feet, the largest of any North American bird, loses some of its majesty viewed up close, when it just looks like a huge black vulture with a bare red head. Eating rotten flesh is not an endearing feature, either, especially when it tends to give the birds and their nests a lingering odor detectable from some distance away.

Egyptian vultures come as close to being attractive as any member of their family, with a bright yellow head that stands out against their tan plumage. They owe the color to carotenoids, but like others of their kind, do not eat any vegetables. Where do they get the raw material for their adornments? Apparently, by eating dung. (Just when you thought it couldn't get any worse for vulture PR.) Several kinds of animals eat dung, since it contains nutrients not absorbed in the digestive tract, but the vultures, which had long been observed to peck enthusiastically at animal droppings in the wild, use the dung as an internal beauty aid. A group of Spanish researchers analyzed the carotenoid content of sheep and cow droppings, and found that while the former were the best source, both provided substantial amounts of lutein, a common carotenoid thought to aid vision in humans. Probably to most people's relief, we humans can get lutein from foods such as leafy greens and egg yolks, but the scientists fed cowpats to four Egyptian vultures in the Jerez Zoo. The birds fell upon the manure with gusto, consuming about three pounds over a ten-day trial period. The dung helped raise the blood levels of the carotenoids in the birds to nearly three times their previous concentrations.

The scientists speculated that vultures seek out dung for, as they put it, "cosmetic purposes." It does not seem to provide any nourishment, and vultures with the brightest yellow heads may be advertising their ability to expose themselves to potential sources of disease in the feces and still maintain their ornaments. According to some of the same Spanish biologists, a related species, the bearded vulture, circumvents this risk by using the vulture equivalent of Clairol: They dust bathe in red soil to dye their feathers red. Lacking this natural boost to their coloration, captive vultures are a pristine white. In the wild, vultures with redder heads are more dominant. The bearded vultures seem to be attracted not only to red dirt but to the color red in general, reputedly being interested in red leaves as well.

Arresting though the image of makeup-wielding vultures is (what would the colors be called? Putrefying Pink? Roadkill Rouge?), it still doesn't tell us whether carotenoid-based beauty signals immunity. The best evidence for that link comes from two other bird species, both of which get their coloration from foods like seeds and insects.

Jonathan Blount and his colleagues from the University of Glasgow studied zebra finches, the tiny spotted birds with red beaks that make that monotonous beeping sound you often hear in pet shop aviaries. The males with the reddest beaks are most attractive to females, and the color is carotenoid-dependent. Supplementing the diet of the males with extra carotenoids not only brightened their bills, it boosted their immune systems.

This means that female zebra finches that choose the reddest-beaked males should also get a mate who can resist disease. Interestingly, if the amount of carotenoids in the diet is decreased for baby birds in the nest, they grow up with lower carotenoid levels in the bloodstream but no apparent difference in their adult coloration or attractiveness compared with zebra finches that got a normal diet while they were young. This suggests that the color of a male's bill is a "what's-happened-lately" indicator of his current condition, rather than a historical record of his childhood.

Blount also pointed out that if female birds do not get sufficient amounts of carotenoids in their diet, they may lay fewer or lower quality eggs. Females in a variety of species have red or yellow ornamental traits, though these are often not as richly hued as their counterparts in males. Researchers have usually assumed that these toned-down sexual ornaments are there because selection on males for elaborate ornaments has inadvertently resulted in females having the same genes, but Blount and coworkers wondered if females could be advertising qualities of their own, namely superior fertility, via the carotenoid-dependent feathers or skin. This is an intriguing idea that remains to be tested.

Even baby animals can exhibit colorful signals using carotenoids. When the barn swallow chicks in a nest open their mouths to be fed, they show off a bright red gape like an arrow pointing to the opening of a trashcan. Nicola Saino and his coworkers found that parents fed the chicks with the reddest mouths more than their paler nestmates. When the chicks were injected with a protein to challenge their immune systems, the color of the gape faded, only to return when carotenoids were administered to the youngsters. It may seem puzzling that the parents did not dole out their affections equally to all the young, since that is supposed to be what parents do. But it is a cruel world out there. All of the baby birds in a nest may not be able

And carotenoids really are special in this regard. Biologists Kevin McGraw and Geoff Hill, both then at Auburn University in Alabama, looked at the effects of an intestinal parasite on American goldfinches, which have yellow feathers and bills that get their color from carotenoids as well as a black cap that uses melanin, the same pigment that gives human skin its hues. The parasite made the yellow, carotenoid-based, traits duller, but it had no effect on the black feathers.

But the mirror does not always reflect the whole story; more detailed studies of the blackbirds' immunity revealed that bill color is connected to some measures of immune response but not others. What's more, whether bill color signaled better immune response depended on which kind of immunity was studied. Recall that the immune system is divided into two arms, cell-mediated and humoral, with the cell-mediated arm using T cells to aid in wound healing and other generalized responses, while the humoral arm depends on B cells and antibodies for a more specific response. The blackbirds with brighter bills had worse humoral but better cell-mediated immunity, which means that the birds might allocate carotenoids into beauty and the two kinds of immunity in a complex way, rather than just divvying them up equally.

To understand how that might work, let us return to my favorite study subjects, the jungle fowl. Unlike domestic chickens, but like many other birds, including the blackbirds studied in France, jungle fowl are seasonal breeders, laying eggs and rearing chicks only during a few months of the year. Before the mating season, males molt into flashier plumage, their combs enlarge, and they become more aggressive, as their testosterone levels mount and they prime themselves for battle. They lose much of their interest in eating, developing a one-track mind for sex and those things connected to it. We were interested in seeing how the males balanced this increased need to look beautiful to females and win fights with other males against the continuing need to maintain their defense against disease, particularly when the females appeared to be choosing mates based on traits that showed off that immune defense.

We measured the birds' cell-mediated and humoral immunity before and during the breeding season, along with their comb size, which we had already determined to be the hallmark of rooster sex appeal. Comb size, incidentally, is correlated with color, and the red of rooster combs partly depends on carotenoid levels. Females in our experiments also preferred to mate with males that had redder combs, though we tended to use comb size as a more unambiguous measure of attractiveness. We discovered that during the breeding season, males with large combs had better cell-mediated immunity but worse humoral immunity than males with larger combs. But outside the performance period, as it were, when sex was not paramount in the lives of the birds, both of the immune measures were better in the large-combed males. My colleague Torgeir Johnsen and I speculated that this kind of selective jettisoning of one part of the immune response makes sense given the extreme pugnacity of the roosters during the breeding season, when the frequency of fights and hence the likelihood of wounding are high. Cell-mediated immunity helps animals deal with cuts and infections, letting them make it to another day and another battle; in the high-risk, high-stakes game male jungle fowl play, that is the aspect of immunity that should be preserved, even if other more long-term responses must be temporarily relinquished. From the female's perspective, though, choosing a male with a big comb still makes sense, both because over the course of the year he is generally more resistant to disease and because her sons are more likely to gamble appropriately.

LOOSENING TESTOSTERONE'S TIES?

Carotenoids are a little like testosterone, in that advertising a trait that depends on either substance is also an indication that you are capable of juggling several needs at once—immune defense versus ornaments with the hormone, foraging versus ornaments with carotenoids. But with testosterone, producing the ornament means taking an obligatory hit in immune defense; increase the level of the hormone, and of necessity also increase the immuno-suppressive effects. Carotenoid-dependent ornaments don't have the same dark side, since eating foods that provide antioxidants actually helps both health and appearance. Eating a diet rich in carotenoids is thus like going to the gym—it can boost your appearance and reduce the risks of disease, and it is under your control rather than that of your genetic makeup. A win-win situation, although of course obtaining the carotenoids in the diet may be difficult if they are scarce out in the wild. Nevertheless, it gives more meaning to the idea that yellow is a cheery color.

Of course, males cannot simply elect to forgo testosterone in favor of red feathers. The hormone is still necessary for many of the basic attributes of masculinity; you can't make a beard with vitamin A, for example, though one might wonder why evolution favored beards in men over, say, yellow patches of skin, like those on the vultures. (Then again, that might have led to the issue of consuming dung, so perhaps we should count our blessings.) Testosterone levels can respond to changes in the amount of circulating carotenoids, too. In mallards, the common duck of parks and ponds, the yellow bill in males became duller when the birds were given an immune challenge, and both testosterone and levels of carotenoids in the bloodstream both plummeted. In addition, the ducks' bill color in autumn predicted the quality of their sperm the following spring, with brighter males producing faster-swimming sperm cells. This suggests that there are complex interactions between male sex hormones and carotenoids, even if the pigments do not themselves impose the same cost as testosterone.

If the only potential cost of carotenoids is obtaining enough to supply both health and beauty needs, why don't males just cheat? Why not reserve what is needed for immune defense and simply fake the rest, using some pigment that can be synthesized more cheaply in the body? Greg Grether, a biologist at UCLA, asked exactly this question in a study of guppies, which have those orange patches on their sides that attract females. A different class of pigment, called pteridines, can also produce an orange hue, but these are simply manufactured de novo by animals, and do not require hustling for just the right seeds, bugs—or dung. Grether determined that, indeed, the orange patches on the fish contain pteridines as well as carotenoids, and proposed that when carotenoids are scarcer in the streams where guppies live, they should shift over to using pteridines. Much to his surprise, they did the exact opposite; in streams with an abundance of carotenoids, the fish incorporated more pteridines, not less, into their skin. Perhaps, he speculated, female guppies are no dummies, and can detect the difference in pigment, like a discerning date who notices a Rolex knockoff.

Finally, before we agree that eating carotenoids is the salvation of animals looking for a way out of the perils of testosterone poisoning, a few caveats are in order. Recall the intricacies of the immune system as I discussed it a few chapters earlier. The pathway between eating carotenoids and becoming healthier is not straightforward, partly because different components of the immune system are often cross-regulated. This means that when one aspect of immunity is very active, another subsides in response, making it virtually impossible to maximize everything about defense against disease. For example, those white blood cells called T cells come in three kinds, and each kind is both useful in protecting against a particular class of ailments and sensitive to a particular nutrient. One kind is important in defense against infections that enter a cell, such as viral diseases, another operates against worms and other larger parasites, and the third regulates the first two types. Zinc and the intake of sufficient calories help support the first kind of T cell population, but are associated with a diminished worm defense. On the other hand, insufficient zinc and calories means that the worm defense flourishes and defense against pathogens inside the cell grows weaker. Vitamin A has the opposite effect, boosting the defense against worms and decreasing cell-entry defense. So are zinc and vitamin A good for you? Yes. Are they also bad for you? Yes. Should you—and the male guppy looking for a hot date as well as a strong immune response—eat foods containing a lot of carotenoids like vitamin A? Well, maybe. Right now that's the uncertain state of the science.

WHO ARE THE BEAUTIFUL PEOPLE?

Enough about vultures and guppies. Are the features that people find beautiful in others good indicators of health? And furthermore, is there a good diet for our own health and beauty? If the answer to these questions is uncertain for animals, it is even more so for humans, although that didn't stop a Web site called Lifescript.com from enthusing, "Phytochemical-rich foods, such as fruits and vegetables, support your beauty and your health, by fighting off free-radical damage that can lead to heart-disease, cancer, and even wrinkles!" Needless to say, the site also sells nutritional supplements. A British company sells pills called "beauty antioxidants" that include carotenoids and are supposed to be better than other antioxidant supplements because they are extracted using a process that takes two weeks rather than the apparently customary fifteen minutes. How a slower process necessarily yields higher-quality results is something my students would love to be able to explain when submitting a late paper.

Carotenoids and other antioxidants in the diet can certainly help us guard against disease. Numerous studies have shown a lower incidence of cancer and other diseases in people who have diets rich in carotenoids, and supplements of beta carotene in particular help the cell-mediated arm of the immune response, especially in elderly people whose immune systems are weaker or who may not always get the appropriate nutrients in their diets.

Whether this leads to greater beauty, much less a signal of a healthier prospective mate, is more questionable. Unlike goldfinches and jungle fowl, humans do not have obvious carotenoid-dependent sexual ornaments like yellow feathers or red wattles. Carotenoids may or may not prevent wrinkles, but from an evolutionary standpoint, it hardly matters, since most of us do not develop wrinkles until we are well into or past our reproductive prime. Particularly among early humans, mate choice, at least the kind that results in offspring, happened relatively early in life, so using a criterion of "unwrinkled" as our sole basis for finding true love would not screen out very many candidates, though it certainly guides us toward the more fertile prospects. Whether fooling a potential mate into believing we are younger than we are is effective is another matter.

Wrinkles aside, smooth, even skin is a great candidate for one of those beauty universals. Preferences for skin colors vary across cultures and in different periods of history, but virtually everyone admires skin that is firm and unblemished. Skin certainly can reveal disease, as evidenced in smallpox, syphilis, scarlet fever, and the bubonic plague, among others. A large part of the horror of smallpox derived from its visible aftereffects, the pockmarks on the skin left by healed pustules that could render the victim blind. Our modern practice of vaccination, in which weakened or killed versions of a pathogen are administered to an individual in order to stimulate an immune response, began as variolation, the practice of giving healthy people a very mild case of smallpox by exposing them to the scabs from pustules. First recorded from Asia, variolation spread to Africa, India, and the Ottoman Empire by 1700, and eventually became popular in Europe. This practice is obviously still risky, since one cannot control the severity of the strain that is administered via the scabs. In 1796, Edward Jenner, an English physician who noticed that milkmaids had exemplary complexions and did not seem to suffer from smallpox, hit on the idea of giving people cowpox, a related but much milder disease common in the milkmaids. This proved successful, and he called the process vaccination, after vaca, the Latin word for cow.

For our purposes here, the noteworthy part of this history is that vaccination seems to have been an effort not only to control the disease, but to eliminate its blemishing effects. A review of the book *Vaccination: The Facts, the Fears, and the Future* by Gordon Ada and David Isaacs notes, "In the seventeenth century, the young beautiful maidens that were to be sold to the Turkish Sultan and the Persian Sophy for their harems were variolated to protect their skin and beauty." Not to keep them healthy, though that is implied, of course, but to keep their skin unmarred. A parasitic disease called cutaneous leishmaniasis prevalent in parts of Latin America, Africa, Asia and the Middle East is not fatal, nor does it cause severe illness, but it is still a substantial public health concern, mainly because the chronic lesions it causes on the face, and chins has to be one of the most unjust rites of passage in the world, and it is the most common skin disease across the globe. Acne isn't dangerous to physical health, but pimples can make people withdrawn, unable to forge strong relationships, and depressed. People with acne are even more likely to be unemployed than people without the blemishes. But in 2004, Dale Bloom argued that acne is actually useful, serving as a signal to prospective mates that the sufferer is too hormonally tempestuous and immature to be a good mate. He points to its conspicuous location on the face and "its ability to elicit reflexive disgust and avoidance in observers" as evidence of acne's ability to communicate information. Pimples thus serve as the opposite of a flashy tail or a red comb, a kind of sexual keep-away signal until the bearer is older and more likely to make a good parent. If you think about it, this too is probably not exactly what teenagers wish to be conveying, but regardless, Bloom's is a novel hypothesis.

While it is plausible to suggest that acne, along with a number of other skin characteristics, is linked to perceived attractiveness, I am not convinced by the notion that it evolved as a signal of sexual unsuitability. For one thing, little or no evidence suggests that modern teenagers with acne are less likely to engage in sexual activity, and for another, in societies where people marry young, the presence of the blemishes is unlikely to serve as a deterrent. The recurrence of acne during times of adulthood well past sexual maturity also argues against the idea. And what about those lucky few who pass through the rough seas of adolescence without the trauma of acne at all? Are they likely to make better parents even at an early age? It seems doubtful. Finally, the real problem with acne for many people lies in the scars that linger after the blemishes subside. These are not confined to adolescents, although I suppose one could test Bloom's hypothesis by seeing whether people only experience that reflexive disgust when confronting fresh cases of acne, and not by skin that merely bears witness to past episodes.

Another idea about skin quality signaling mate attractiveness was tested by Bernhard Fink and Karl Grammer from Vienna and Randy Thornhill of the University of New Mexico. They used a computer program to make composites of photographs of women's faces, and then created faces that had the same shape, but differed in skin color and texture. They then recruited men to look at the images and rate their attractiveness. Smoother, more homogeneous skin was ranked higher than coarser skin, and darker skin tones were preferred over lighter ones. The scientists suggest that skin texture and color are part of a group of characteristics, including the symmetry of facial features on the right and left sides, generally perceived as beautiful.

A group of Scottish psychologists generated photographs of skin patches, from men this time, and asked women to judge how healthy a person with each of the skin types was likely to be. These ratings were then compared with the attractiveness scores the women gave to photographs of faces with the different skin types. In a different experiment, the same psychologists asked participants to rate the attractiveness of images of the actual faces, with the color and texture of the skin digitally manipulated to suggest good, average, or poor apparent skin health.

Not particularly surprisingly, women tended to judge the photos of men as being more attractive if they had better skin, and they were easily able to rank the skin samples as more or less healthy in appearance. But while healthy-looking skin made the women rank a photo as more attractive than the same face manipulated to have coarser or more uneven-textured skin, the tendency was weak, begging the question of whether skin is a deal-breaker in mate choice.

Even if we do use skin to signal attractiveness, we still don't know whether a nice complexion means a healthier mate. With the roosters and guppies, it was relatively easy to inject them with a foreign protein, assay their immunity, and measure the effect on both their sexual ornaments and their likelihood of being chosen as a mate. Try doing that with even the most willing of students in an introductory college psychology class, the favored subjects for much of the research reported here.

DROP-DEAD GORGEOUS:
THE ERRANT SKIN OF HISTORY

At least some of the time, people have been led badly astray by attending to the effects of disease on skin. Two historically important diseases are caused by very similar bacteria, roughly as much like each other as dogs and coyotes, or mallards and black ducks. Although both are serious illnesses, one is relatively easy to transmit, while to catch the other requires sustained contact with an infected individual. The first is also hard to cure, while a few doses of the appropriate drug, given at an early stage, completely eradicates the other. One of the diseases has been met with repugnance, with its victims seen as bearing the wrath of God, while the other was romanticized and those suffering from it thought to be creative geniuses. Because of the different effects of the diseases on the skin, the highly contagious disease was the one imbued with glamour and the one that is hard to catch was shunned. The contagious disease is tuberculosis. The other, leprosy.

Today tuberculosis does not have much cachet, but in the Victorian period it conjured up the image of a fevered artist, romantically slender, with bright eyes and flushed cheeks, producing great works and dying at a tragic young age. It is transmitted readily when the blood and sputum coughed up by infected individuals contacts others, but this risk was not recognized until relatively recently. Also called consumption because of the way that the disease seemed to devour its victims, tuberculosis struck down many of the great heroes and heroines of both history and literature, including Puccini's Mimi in *La Boheme* and Thomas Mann's Hans Castorp in *The Magic Mountain* as well as Jane Austen and Edgar Allan Poe. People believed that the disease itself conferred some kind of creative ability on its victims, making geniuses out of ordinary people and transforming geniuses into towering heroes. The later stages of tuberculosis are in fact associated with a kind of euphoria, with dying sufferers often talking excitedly about plans for the future, much to the consternation of their friends and relatives, which may be where some of this image developed. In addition, the weight loss and subsequent thinness, pallor, and lethargy caused by tuberculosis were all desirable attributes of the upper classes at the time. In an odd twist, the blood that is coughed up by advanced tuberculosis victims was also viewed favorably, because blood was a fascinating substance to the Victorians, tied up with both reproduction, through menstruation, and death.

In contrast, leprosy is probably one of the oldest diseases with a social stigma, with victims shunned by society and held in leper colonies or forced from their homes. Ironically, such isolation would have been of far greater use in the control of tuberculosis, because the bacteria that cause leprosy grow extremely slowly and there is no transmission route comparable to the infected blood and saliva produced by those with tuberculosis. Many people also seem to have an innate resistance to leprosy, which is why relatively few of the religious figures working in leper colonies actually contracted the disease; one study of over 22,000 people in contact with untreated leprosy patients, including family members and other caregivers, showed transmission in less than 1 percent of the cases. But leprosy (now usually called Hansen's disease, after the scientist who isolated the bacteria that causes it) has a problem that tuberculosis doesn't: It disfigures the skin and appearance of its victims. Even the word "leper" reputedly originates from a Greek term for scaly, and because of the nerve damage to extremities caused by the bacteria, sufferers frequently have missing fingers or toes or other skin damage because they do not protect themselves from injury. The affected parts don't "fall off" the way the jokes and horror stories have it—the tissue is absorbed as part of the infection. Although prejudice against those with the disease remains, modern drugs have helped to control it in many parts of the world, and it is not a major threat to our survival. Tuberculosis, on the other hand, while at an all-time low in the United States, is still a threat in much of the Third World, and the evolution of drug-resistant forms of the disease poses a major public health problem.

There is no need to dwell on how misguided our responses to illnesses have been and continue to be, save to point out that the tuberculosis-leprosy contrast shows all too clearly that our instincts about disease avoidance are often flawed. This huge error in judgment also suggests that our ability to rely on skin condition to tell us about who is likely to make us sick, or to have susceptible offspring should we choose them as a mate, is imperfect at best.

Other attempts to find a cue that humans use to indicate health have suggested hair as an alternative to skin; Hungarian psychologists Norbert Mesko and Tamas Bereczkei presented photographs of female college students with and without different hairstyles (long, short, medium length, "disheveled," "unkempt"—it isn't clear how these last two differed—and a bun or knot) to groups of young men. The researchers suggested that if people choose hair that indicates the presence of parasites or disease, the men should find the faces with unkempt or disheveled hair less attractive. Perhaps the concept of "bed head" being sexy has not made it to Hungary. In any event, the only style that seemed to appreciably increase attractiveness was long hair, and this also increased the perception that the woman in the photograph was healthy. Another study involved two researchers interviewing women in parks and college campuses of the Upper Midwest in the United States; one of the researchers asked the subjects questions about their age, health, and other information, including their marital status and number of children, while the other researcher lurked nearby and noted attributes of the woman's hair. Indeed, women who claimed to be in better health were judged to have slightly better quality hair by the observers, although the correlation was not particularly strong.

These findings do not say much about the value of hair as an indicator of disease resistance, in my opinion. Hair fashions are extremely variable across cultures and throughout history; a group of the ever-willing college students would probably rank photographs of people with seventeenth-century hairstyles as less attractive than those with modern coiffures. But this does not mean that the difference in styles between the eras occurred because of selection for healthier mates. Indeed, a group of seventeenth-century subjects would probably prefer photos of women with the appropriate hairstyles for their era too, fleas and all. Unkempt hair may well be connected with ill health, because those who are sick are less interested in personal grooming, but that is a trivial result, and not one that helps us see how traits such as skin or hair could have evolved specifically to show off health, like finch beaks or rooster combs. And hair may be a more complex signaling device than the authors of the studies realize; any issue of *Elle, Vogue,* or *Glamour* is likely to have a two-page spread on the intricacies of different kinds of bangs alone. Short versus long doesn't even begin to describe our attention to detail.

For that matter, the same may be true of our evaluations of skin, with the cosmetics ads and magazine articles exclaiming over a shift from pink to peach in cheek color (pink, according to one makeup guru, is too "girly," whereas peach is "fresh").

THE BEAUTY MYTH OF BLIND LOVE

This is not to deny that scientists are discovering some universally attractive features of the face and body. Female faces with high, prominent cheekbones, male faces with longer lower jaws, and female figures with a waist-to-hip ratio of .71 (achieved, for example, with a waist of 25 inches and hips measuring 35 inches, or a 28-inch waist and 39-inch hips) have been found to be appealing in a number of studies. These characteristics might be linked to fertility, because they can reflect the levels of sex hormones like estrogen or testosterone. Both symmetry of features and facial "averageness," or lack of extreme features, have also been found to be attractive to people. But the question is whether any aspect of attractiveness, whether it is skin, hair, or some combination of these factors, truly indicates health, not just perceived health or imagined health.

Two sets of researchers set out to address exactly this question. The first group, led by S. Michael Kalick of the University of Massachusetts, used the ubiquitous photos again, taken in late adolescence, and they asked raters to rank the attractiveness of the individual pictured. A different set of people was then asked to rate the apparent health of the subjects in the photos. The clever twist was that the researchers obtained actual health data on the people in the photos, using health records starting when they were teenagers and continuing into late adulthood. Adolescents were scored on the frequency and severity of illnesses ranging from colds to measles, while adults' health was based on an exam and medical history obtained from physicians. This meant that the scientists could verify not only whether people thought attractive individuals were healthy, but—more important—whether they were right.

As would be expected if beauty does indeed indicate health, being rated high by the people giving scores for attractiveness meant that an individual was also likely to be rated high by those ranking health. That perceived level of health, however, turned out to be a poor cue of actual well-being. The raters did a reasonable job at accurately gauging adolescent health for those individuals with medium levels of attractiveness, but when the people in the photographs were judged to be very attractive, this ability went out the window. In other words, being beautiful is distracting to the observer. Instead of being an indicator of health, attractiveness may be a decoy, or, in more technical terms, attractiveness functioned as a suppressor variable.

The second study asked observers to look at photographs of faces and rate them on a 1–7 scale of "masculinity" (for male faces) or "femininity" (for female faces), and then to score the photos for perceived health as well as attractiveness. The researchers, from the University of Western Australia and Brandeis University in the U.S., then obtained health information from medical examinations and histories, and as in the previous work, looked for any concordance between how healthy the observers thought an individual was, how attractive they thought the person was, and how healthy that individual was in reality. More feminine female faces were seen as both more attractive and healthier, but in fact the individuals with those feminine features were no healthier than any others. Male faces that ranked high on the masculinity scale were perceived as healthy, though not very attractive. Again, no foolproof window to the immune system seemed to be forthcoming.

Maybe one of the reasons that we can't seem to find a reliable indicator of health in an attractive face or body is that our ideas about attractiveness are more than physical. It seems plausible that a good sense of humor, intelligence, or compassion will outweigh a flawless complexion, impressive set of biceps, or perfect waist-to-hip ratio. Or are we deluding ourselves? Is this kind of compensation just the fond hope of the less-than-beautiful looking for consolation?

This notion has been examined by many psychologists, but an anthropologist and an evolutionary biologist from Binghamton State University in New York recently provided an exceptionally interesting analysis of it. Kevin Kniffin and David Sloan Wilson wanted to know whether personal experience with someone altered the perception of that person's attractiveness. First, they asked participants to rate the ever-popular photographs, but instead of asking for a dispassionate evaluation of images that could have come from a magazine, they used photos from the subjects' high school yearbook. A stranger of the same sex and approximately the same age as the subject then rated the same photos. Although the strangers' and the classmates' assessments were similar, the amount of respect, liking, and familiarity felt by the classmates made a big difference in the rankings. In other words, knowing how much a subject liked the person in the photo was as good a way to predict whether the subject would think that person was attractive as knowing whether a stranger had thought he or she was attractive.

Kniffen and Wilson then extended their study by having Kniffen infiltrate a university rowing team for a season. At the end of it, he asked the members to rate their teammates for "talent, effort, respect, liking, and physical attractiveness." Strangers were then asked to rank photos of the rowers for attractiveness. Once again, liking and respecting a person strongly influenced how team members ranked that person's attractiveness, with the strangers' evaluations jibing rather poorly with the team members' ratings. Finally, the researchers queried members of a six-week summer archaeology course about the qualities of their classmates before and after they had been through the camaraderie of life in the literal trenches. As you can probably guess by now, it was hard to predict the final attractiveness rating based only on the initial impression; one woman whose attractiveness was ranked at 3.25 on the 9-point scale at first turned out to be well-liked and a hard worker, with a final score of 7. Beauty was as beauty did, or at least as it rowed, dug for artifacts, or helped others with their homework.

BEAUTY'S BETTER NATURE

The failure to find a single human feature, whether skin, hair, or symmetry, that shows off health and guarantees allure should not come as a surprise. Anyone who really believes that we choose our mates as if we were bowerbirds, with a snap judgment based on a single, health-revealing trait, must not get out much. We have at our disposal a host of cues to an individual's quality, and what's more, we need more from our mates than the bowerbirds do. A female bowerbird sees a male for a few minutes or perhaps hours, mates with him, and leaves.

In contrast, humans are at least ostensibly in a relationship for the long haul, and it pays to use all the information we can get about a variety of important attributes. We do not need to use skin or hair the way that hens use rooster combs or female bowerbirds use bowers. The people looking at photographs were rating strangers' looks, not getting married to them. Love at first sight may be real for some people under some circumstances, but we do not hear about the times that our first impressions led us astray and what we thought was love was just an attractive illusion.

It is reasonable to think that our ideas of what is beautiful were shaped by the advantages of choosing a disease-resistant mate. But it is absurd to go on to conclude that we should adhere slavishly to a single cue, whether it is skin or symmetry, as a health indicator, when we have so much more at our disposal. If a potential mate is coughing, weak, and vomiting, should we resolutely stick to an examination of his or her blemishes? Attraction is mysterious, and that adorable pancreas is an elusive goal. Health, rather than truth, may be contained in beauty. But no single feature will unlock its secrets.

CHAPTER  9

TAKING CARE

Amidst all the panic over Ebola, West Nile virus, and SARS, as we research new treatments and plan our attack, an important question goes unasked. Why aren't we sicker than we are? We move unwittingly through a sea of swarming microbes and writhing worms, and even if many of them have their good points, as I discussed in previous chapters, that still leaves plenty of nasty diseases to go around. Yet the sun rises and sets, day after day, on a monotony of well being, with most of us, even in countries without good hygiene and modern medicine, feeling reasonably healthy. The same is true for other animals; although parasites are certainly a continual threat, many if not most individuals manage to avoid succumbing. Why don't the pathogens win even more often?

Part of the answer, of course, is our exquisitely complex immune system. Our bodies have evolved to recognize and dispatch foreign substances with alacrity. But this physiological response is not the whole story. Humans and nonhumans alike do not simply wait passively for the next infection to come along and then lie back and hope their lymphocytes are not sleeping on the job. Instead, we all have a vast array of behaviors that we use to keep from being infected in the first place. Disease prevention is not just an outdated high school health lecture; it is part of what blurs the line between food and medicine, between grooming ourselves out of vanity and doing so to keep parasites at bay. If your cat stops licking and scratching itself, it won't just have a bad fur day, it will risk developing an uncontrollable population of fleas that not only make its life miserable but can transmit a variety of nasty pathogens.

Our own actions can keep us from getting sick, or if that doesn't work, at least keep us from getting any sicker. Behaviors as disparate as our cooking practices and our squeamishness around scatological topics may have arisen because of disease, long before consciousness itself arose. We take echinacea and vitamin C to ward off colds, or caution our children not to go barefoot or pick up dropped cookies lest they harbor germs, and we feel civilized, sophisticated, advanced beyond our animal roots. But we fail to note two things. First, many of our disease avoidance behaviors date back well before the discovery of germs, to our earliest ancestors, and they may be things we do without ever being aware of them. Second, many other species, from ants to parrots to monkeys, thwart parasites and stay well without ever realizing the perils lurking in public toilets. What do we—and they—do, and how did we all figure it out?

THE DIRTY DIET

What do monkeys in Puerto Rico, macaws in Peru, pilgrims to an ancient church in Chimayo, New Mexico, and the Luo people of western Kenya have in common? They all eat dirt. They do so not just incidentally because a piece of fruit drops on the ground, or because they are having a picnic outdoors, but deliberately, in big gritty mouthfuls. Humans in many societies, both traditional and modern, regularly eat dirt, and the usual medical response is that a little bit is okay, but consuming larger amounts—defined in

2000 by the U.S. Agency for Toxic Substances and Disease Registry as more than 500 milligrams, or about half a teaspoon—is potentially worrisome. (How this threshold was hit upon is unexplained.) The craving for and consumption of nonfood items like dirt, ice, or paint is called pica, after the scientific name for magpies, birds that sometimes accumulate a variety of objects in their nests. It is considered a disorder if it persists for longer than a month, and it has been linked to obsessive-compulsive disorder and eating disorders as well as schizophrenia in humans.

But for the Luo, earth-eating, or geophagy, the more technical term, is a practice filled with spiritual significance. It is most common among prepubescent children and women of childbearing age, and it is linked to the female side of nature, so that the ingested earth "symbolizes female, life-bringing forces," according to anthropologist Paul Wenzel Geissler. To the pilgrims who come to Chimayo, the earth has been blessed by Christ, and can work miracles.

Geophagy has been viewed by physicians as a pathological behavior partly because it seems inexplicable and partly because it can sometimes cause damage to the intestinal tract. The earth can also bring worm eggs along with spiritual meaning. Geissler and his colleagues interviewed 204 Kenyan schoolchildren and found that 77 percent of them ate soil daily. Many of the children harbored worms, which was not unexpected, but after they were treated for the parasites, reinfection was twice as common among the children who ate dirt as it was among those who did not. Pregnant and nursing women in Kenya were also more likely to reinfect themselves with parasites, particularly roundworms, if they routinely ate dirt.

But wait—this chapter was supposed to be about avoiding disease, not participating in bizarre behavior that increases it. Indeed it is, but to understand why eating dirt is part of our on-going dance with disease, we first need to look at the other species besides humans that engage in the same behavior.

Ecotourists to South America are eager to see large, colorful animals, and their guides are equally eager to oblige. Some of the most beautiful inhabitants of the Amazonian rain forest are the macaws and other gaudy parrots, but seeing one of these birds on the wing overhead is often disappointing; silhouetted against the sky, or in a blur of foliage high in the canopy, a bright red, green, and yellow bird can look surprisingly dull, and they fly so quickly that they often disappear from sight before binoculars can be trained on them. But the guides often know places where hun-dreds of parrots festoon riverbanks or cliff sides, like so many Christmas ornaments. The birds, who stay pressed against the banks long enough to allow leisurely photography, are not eating their usual fruit, or nesting, or even posing for tourists. They are eating a specific type of dirt found along the river.

The earth they consume, like the dirt eaten by many of the human groups that practice geophagy, is high in clay. Parrots eat many fruits that are high in alkaloids and other nasty plant tox-ins. Clay binds to the noxious chemicals in the gastrointestinal tracts of the birds, and renders it harmless. Dirt-eating is the par-rots' way of circumventing the plant defenses, enabling them to eat a wider range of foods.

One of the constituents in the clay is kaolin, similar to the active ingredient in Kaopectate and other commercial anti-diarrhea medicines. Although the parrots do not generally suffer from worms, since those parasites are usually transmitted via fecal contamination of soil that contacts food, other animals that eat kaolinitic clays do. Among a group of rhesus monkeys, for example, 89 percent of the 141 members were infected with parasites, but diarrhea was virtually absent, even though diarrhea is a common symptom of worm infection. More than three-quarters of the monkeys ate dirt high in kaolin, and Mary Knezevich, who studied the rhesus troop, along with other scientists, suspected that regular geophagy in the rhesus monkeys as well as other species helps control symptoms that would ordinarily accompany the levels of parasites carried by many wild animals.

Getting back to the supposed pathology of pica, geophagy has also been suggested to supply minerals not ordinarily obtained in food, a kind of natural dietary supplement, which might explain its pervasiveness among pregnant women, who can become nutritionally deficient as nutrients are shunted to the developing baby. At other times, it may have evolved simply as a way to ameliorate the effects of pathogens. One enterprising study of the potential benefits of soil consumption used a model intestine, made of glass and plastic and complete with mock digestive fluids, to discover that kaolin clay could easily absorb toxins such as quinine and tannins.

I doubt that earth-eating, particularly as practiced by nonhumans or by large groups of people like the Luo, is a sign of psychological disturbance, unless you think that most macaws are mentally unbalanced. The increased level of reinfection with worms that the earth-eating Kenyans experience may be an artifact of modern life; perhaps in earlier times people moved around enough that worm eggs did not have an opportunity to build up in the soil that was eaten. Or alternatively, perhaps maintaining a certain level of infection with worms was tolerable so long as the symptoms were managed by eating the soil. Remember, the idea that we—or any other organism—should be completely parasite-free is a modern invention, not a natural state of affairs. Effective deworming would have been unthinkable during the time when we were evolving with our diseases, and the benefits of eating dirt outweighed the harm of a worm egg or two.

DARWIN MADE ME EAT THAT

If consuming dirt, even to neutralize the effects of parasites, is unappealing, how about eating a bowl of chili? Paul Sherman, an evolutionary biologist at Cornell University, suggested several years ago that the use of strong spices in foods like chili or the curries of Asia does more than add flavor, and might be the result of more than a mere cultural lottery. According to Sherman and his colleagues, humans evolved to use spices as antimicrobials, killing the bacteria and fungi that are particularly likely to contaminate foods in regions like southern India and Brazil, where the climate is warm and microorganisms thrive.

Although spices are used throughout the world in cooking, we all know that the number and kind of spices varies greatly; no one goes to a Norwegian restaurant for a highly spiced meal. Sherman hypothesized that if spice use evolved to deter bacterial contamination, traditional recipes from countries where the risk of food spoilage is high should use more spices. Furthermore, the spices themselves should actually work to kill or inhibit microorganisms associated with rotten food, and the spices that are the best at this task should be used the most in geographic regions where they are particularly needed. He and his student Jennifer Billing also examined various alternative explanations for regional variation in spice use, including the idea that spices mask rotten flavors or odors, or that eating spicy foods helps people cool off by encouraging sweating. They ultimately rejected both of these, noting that going ahead and eating food contaminated by illness-causing microorganisms should be discouraged, not favored, by evolution, and that while eating chili peppers causes increased perspiration in some people, the effect is far from universal. In addition, cooling oneself through sweating is a rather expensive proposition, metabolically speaking, and it seems unlikely that it would be a viable tool for coping with a steamy climate year-round.

Sherman and Billing ended up using—though, alas, not actually preparing—4,578 recipes from about 100 cookbooks with traditional recipes of 36 countries ranging from Norway to Indonesia to Hungary. They also obtained information on the mean annual temperatures and rainfall for each country, and they investigated the efficacy of the spices most commonly used at killing or inhibiting growth of pathogenic bacteria. As they had predicted, recipes from hotter countries used more spices, particularly those spices that were especially good at keeping bacteria from multiplying: chili peppers, garlic, and onion (which they classified as spices rather than vegetables because of the way they were used), cumin, cloves, lemongrass, oregano, and cinnamon. Recipes from the two coldest countries, Finland and Norway, often called for no spices at all, while in the cookbooks of sultry India, Indonesia, Malaysia, Nigeria, and Thailand, every single recipe with meat in it used at least one spice, and many of them called for ten or more.

The initial work focused on meat-based recipes, since the microbes that infest meat are more dangerous to health than those that might grow on vegetables. Vegetables also often have an inherent ability to fight bacteria, probably because they contain the same chemicals that make spices themselves effective antimicrobials. The acidity of many plant cells is also too high for many bacteria to grow very well.

A later study showed that even in countries like India where many spices are used in food, vegetable dishes traditionally contain fewer spices than meat dishes. And it's not just that spices are more available in hotter countries; spices often travel long distances regardless of the country where they are eventually used. Spices have been valuable to people from virtually all cultures for millennia, which probably explains why they are not used everywhere, including places with a lower risk of foodborne illness; they are simply too expensive to squander if not needed to prevent infection. Interestingly, two of the most universally used seasonings, black or white pepper and lemon juice, may help make the active ingredients in other spices even more potent.

Billing and Sherman also compared rates of food poisoning in Korea, which traditionally uses many spices, and Japan, which does not. Between 1971 and 1990, food poisoning affected nearly three times as many Japanese as Koreans. The authors suggest that in early Japan, fresh seafood was always available, so the need for spices did not arise, whereas in more modern times, as food began to be transported over long distances, the opportunities for spoilage increased.

Food scientists have explored the antimicrobial activity of spices, and generally supported Sherman's hypothesis. An elaborate study of garlic in chicken sausage found that the flavoring extended the shelf life of the meat product longer than a standard chemical preservative, BHA.

GUT FEELINGS

Needless to say, neither Sherman nor anyone else suggests that early humans hit upon these mechanisms for avoiding illness consciously ("Gosh, a little clove would really help kill the bacteria on that mammoth you just brought into the cave"), or deny that the immediate reason for using spices is that people like their flavor and aroma, or that cooking improves flavor. But saying people prefer a certain flavor begs the question of where and why that preference arose, and the answer seems to be that it is at least in part an adaptation against pathogens. I think that Sherman's findings argue, once again, for our deep heritage of dealing with disease.

Cooking food helps kill parasites, too, along with neutralizing toxins in plants, and at least one ethnobotanist, Timothy Johns, suggests that heating food may have evolved as a means for humans to deal with those toxins and thus broaden our diets. I wonder how much reducing the transmission of worms may have played a role as well; many people are familiar with the risks inherent in eating undercooked pork, which can carry encysted worms, but fewer seem to recognize that eating sushi, ceviche, and other meals containing raw animal flesh also carries risks. Lemon juice and spices may be effective antibacterials, but they have little or no effect on multicellular organisms like worms. A friend of mine once found a worm waving insouciantly from the bottom of a bowl of ceviche served at one of the nicer restaurants in San Francisco. It stayed alive long enough to be taken into the laboratory and identified as a human pathogen.

Many of those who recoil from using a public restroom for fear of infection think nothing of downing raw fish in these trendy foods. There is something chic yet atavistic about consuming uncooked flesh attractively presented. Yet the shining slabs of seafood have a dark underbelly. I have a T-shirt from the Northern California Society of Parasitologists, a group of scientists who gather to talk shop and show admittedly grisly photos. The front says, "If you knew sushi like I know sushi . . ." On the back is the life cycle of a worm, *Anisakis,* that finds its way into fish, especially marine species like cod, halibut, mackerel, and herring, and thence—via those elegant morsels of rice, nori, and seafood—into people.

Anisakis is not supposed to end up in the sushi bar, much less in you. Humans are what are termed "accidental hosts." Ordinarily, the worm wends its way from an egg in the ocean to a tiny swimming baby worm that is eaten by a crustacean like a shrimp. The worm-containing shrimp is then consumed by a fish, perhaps a herring. The fish in turn, still harboring the worm, either has the good luck, at least from the worm's perspective, to be eaten by a marine mammal such as a dolphin, or the bad luck to find itself on ice under a chef's knife. The worms are often invisible, so inspections are not particularly effective, and fresher fish is no less likely to harbor a parasite. If you ingest the worm, it can cause severe intestinal and gastric disturbance or an allergic reaction. Prolonged extreme freezing can kill the worms, but will not necessarily remove the risk of the allergic response to the worm proteins.

With the increased popularity of sushi worldwide, infections with *Anisakis* and related worms are also on the rise; the majority of cases, about 2,000 per year, are from Japan, but the U.S. and Europe come in for their share as well through the consumption not only of sushi but of pickled herring, cured salmon, and other uncooked fish. The parasite has to be removed surgically or via a tube inserted in the throat, since drugs are largely ineffectual, although it is also likely that in some cases the worm disappears quietly without causing symptoms.

So should you avoid sushi? The answer depends on your liking for both the food itself and for avoiding risk. The chance of an infection is quite small—only about fifty cases per year are reported in the United States—so modest consumption is very unlikely to make one ill. At the same time, I must confess that I do not eat raw fish myself, partly because I was convinced not to after a dinner with a group of parasitologists, a generally skeptical breed of scientists not given to squeamish or alarmist behavior. The parasitologists simply did not think it was worth the chance, and I decided to honor their prudence.

But squirmy stories aside, *Anisakis* is a problem for people precisely because of our history of evolution with disease. We are not the worms' correct hosts; they complete their life cycle with much less disruption in the gastrointestinal tract of a porpoise or seal. It does the worms no good to end up in our guts, and the allergic reaction they induce is a highly unusual response to such a parasite.

What about other worms? In 1946, Norman R. Stoll gave his presidential address to the American Society of Parasitologists, which was meeting in December in Boston. Titled "This Wormy World," it surveyed global infections with parasitic worms of various types, and it has become a classic, updated in 1999 by another renowned parasitologist, D. W. T. Crompton. The conclusion of both scientists was that the world was pretty wormy, whether viewed from a post–World War II vantage point or a relatively recent one, but they differ markedly in the geographic focus of their concerns. While most cases of parasite infections have always centered on tropical and subtropical regions, Stoll was alarmed by the prevalence in the United States of trichinosis, a worm present in many kinds of mammals that infects humans when they eat meat, particularly pork, harboring the worm's eggs or larvae. At the time of his address, which is really not so long ago, about one in six Americans were infected with the worms, a truly astounding proportion for a supposedly modern and hygienic society. Stoll railed against the then-common practice of feeding garbage containing meat scraps to pigs later slaughtered for the table. He acerbically notes, "Those of you who travel the Pennsylvania R.R. into New York become aware that we maintain in the Jersey meadows at Secaucus a malodorous demonstration for every traveler to observe how it is done. When we Americans manifest an unbecoming impatience at how slothful other peoples are, in undertaking the necessary and obvious steps to free themselves from this, that, or the other endemic helminthiasis [helminth is a general term for a parasitic worm], let us pause a moment with ourselves."

Thanks to much improved programs of meat inspection and pig farming, the incidence of trichinosis has decreased dramatically in the United States, and is now at less than 1 percent in commercial pigs. It is worth remembering, though, that this happily wormless state of existence is precarious. The parasite is still a problem in other parts of the world, and a 2003 study noted that it has reemerged in troublesome numbers in Serbia, probably because the dissolution of the former Federation of Yugoslavia led to accompanying breakdowns in agricultural inspection and meat processing. Pausing a moment with ourselves from time to time to consider intestinal parasites might not be a bad idea.

Other currently serious parasites of humans include hookworms, gut parasites that when examined under the microscope bear a startling resemblance to the voracious invader in the *Alien* movies; an estimated 1,298 million people worldwide, generally in developing countries, are infected. Whipworms, blood flukes, and roundworms are also common in many parts of the Third World, and more than one third of the world's population has one kind of worm or another. These days, unlike the post–World War II period of Stoll's concern, cases are mainly confined to developing nations. Virtually all of these infections are due, in Stoll's words, to man's "ineffective insulation from his own excretory products." This means that improved public sanitation, including dealing appropriately with food sources, would go a long way toward decreasing the problem. In most cases, the effects of infection are exacerbated by poor nutrition; an individual with a good diet and a low number of worms does not generally have many ill effects, though that is scant consolation to those from Western societies who pick up a worm infection while traveling.

FRAGRANT NESTS, AND WHY DOGS ROLL IN GARBAGE

Birds' nests are inherently appealing, mimicking as they do tiny homes in the woods or fields, painstakingly constructed out of twigs or mud or leaves. Biologists have studied nest-building for many years, with collections of samples from different species forming the start of many ornithologists' careers. But in addition to the usual stems and other dried material, many birds will carefully harvest and insert fresh leaves, usually from aromatic plants like fleabane and wild carrot, into the nest matrix. Scientists suggested functions for this behavior ranging from the aesthetic (the greenery adds just the right touch for the avian Martha Stewart) to the functional (green material helps shade the eggs, or adds to the humidity in the nest).

The aromatic plants often used suggested to Larry Clark, a scientist now at the United States Department of Agriculture, that birds might actually be fumigating their nests to keep fleas, mites, and other ectoparasites at bay. These small bloodsucking animals can be serious pests, draining substantial amounts of blood from vulnerable nestlings and causing anemia, stunted growth, and even death.

Clark performed a simple experiment using European starlings, birds that nest in tree cavities or manufactured nest boxes and that commonly place green vegetation in their nests. Starlings were introduced into the United States in the nineteenth century and are now common virtually everywhere in the country. He removed the nest material from one set of nest boxes and replaced it with dry grass. Another set of nest boxes received both the dry grass and a handful of wild carrot leaves woven into the nest, with the greenery replaced at five-day intervals. The starlings often use wild carrot in their nests, and they accepted both types of nest readily. Then Clark counted the number of bloodsucking mites in each nest type.

By the time the chicks had left the nests, the nests with the wild carrot had an average of 3,000 mites, while the nests without the fresh greens had 80,000, a nearly 27-fold difference. Work on blue tits, European birds related to chickadees, showed a similar use of fragrant plants such as lavender and yarrow, gathered with eight or so other herbs in a veritable potpourri, with the birds themselves replenishing their supplies after a day or two as the aromas began to fade. People have used the same plants for centuries as ingredients in cleaners and disinfectants.

Nest fumigation also suggests that birds may have a much better sense of smell than had previously been supposed, since they must be able to tell which plants to choose and when to replace them. The olfactory-related portion of the brain in birds is relatively small, and people had assumed they did not pay attention to odors in the environment. If these findings are general to many other species, this assumption may have to be rethought.

Another group of researchers examined the effects of nest plants on bacteria that degrade feathers. Such bacteria can exist in large numbers on adult birds, and scientists are only now discovering their potential for nibbling away at the structural integrity of the feather. The kinds of plants found in nests contained several chemicals, including ascorbic acid, that significantly reduced the growth of the harmful bacteria.

Wood rats, sometimes called pack rats after their predilection, like magpies, for collecting bits and pieces of material to store in their nests of sticks, also place fresh leaves of the California bay on or near their sleeping places. They will nibble the edges of the leaves before adorning their nests, a practice that releases the volatile chemicals inside. Richard Hemmes from Vassar College, along with Arlene Alvarado and Ben Hart from the University of California at Davis, placed bay leaves, torn to simulate the chewing by the wood rats, in jars with fleas. After 72 hours only 26 percent of the fleas survived, compared with 87–94 percent left in jars with other leaves, such as oak, not selected by the wood rats.

A rather novel method of clearing nests of parasites is employed by eastern screech owls. These small owls will nest in boxes nailed to trees by investigators, similar to the bluebird nest boxes placed in many parts of North America by songbird lovers. Fred Gehlbach of Baylor University in Texas has been studying the owls for nearly forty years, and he and his colleagues have repeatedly seen an odd sight: An owl with nestlings will fly back to the box clutching a live blind snake in its talons. The owls do eat snakes, but they kill their other prey before it is taken to the nest. The blind snakes, however, are simply left at the bottom of the nest, where they survive for some time, eating insect larvae that prosper in the debris of rotting food and that may otherwise end up parasitizing the young owlets. Gehlbach suggests that the insects would either compete with the nestlings for food or else suck blood from the nestlings, which would explain his peculiar discovery that the nestlings from nests with snakes grew faster than those without their reptilian companions. How the owls manage to be such discriminating herpetologists, selecting only the appropriate species, and why they only do this on occasion, remains a mystery.

Animals without nests, or birds outside of the breeding season, may still employ parasite repellants. A behavior rather elegantly called self-anointing, in which an animal rubs material over its body, has been observed in birds, monkeys, and several other mammals. In capuchins, also known as organ grinder monkeys, the animals gather in small groups and go into a frenzy of drooling and chattering as they rub material from four different plants, including a kind of native citrus fruit, over their fur. Mary Baker, the anthropologist who discovered this phenomenon, says that the monkeys seem to be in an orgy of excitement during these bouts, and videos of the capuchins show them carefully selecting the plants they will use and then retreating up into the treetops to indulge. The plants that are used seem to have insecticidal properties, and some are active against bacteria as well. Baker tried using the citrus fruit on her own skin; unlike commercial citrus like oranges or grapefruit, the juice from the wild fruits was not sticky, and left her mosquito-bitten arms feeling soothed and refreshed. Cosmetic companies, take note, though you might want to offer a caution about the drooling.

Dog owners who drag their pets away from rolling in rotten fish—or worse—may find this behavior less objectionable, or at least more explicable, once they realize that some biologists classify this behavior as self-anointing as well, with wildlife biologist Paul Weldon referring to "canids that roll in decomposition products," a phrase that makes it seem almost sophisticated. It is not clear why smelling like putrid flesh is advantageous to the dogs, or whether it deters fleas or other parasites, but this is certainly an avenue worth investigating.

Other animals use insects or millipedes for self-anointing; anyone who has had occasion to pick up one of the many-legged arthropods may have noticed that they sometimes produce a distinctive sharp odor. The active ingredient, from a class of chemicals called benzoquinones, is toxic to at least one kind of tick found on the monkeys that rub the millipedes over their fur and skin. Anting, a behavior in which birds either place ants in their feathers or lie on the ground with wings outspread, allowing ants to crawl into the feathers, has long been suggested to be a form of personal fumigation using formic acid and other noxious chemicals produced by ants. Anting has been documented for many years, with most longtime bird watchers having an anecdote or two about seeing a blue jay or other songbird lying near an ant trail or nest in a peculiar position or carefully picking up ants and putting them under the wings. Surprisingly, however, a careful study of the ability of ant-produced chemicals to kill bacteria and fungi showed no effect of formic acid in the concentrations usually released by the ants. Even a suspension of ground-up ants in water didn't kill the microbes, although a purified form of formic acid, much stronger than that found in the ants' bodies, did inhibit bacterial and fungal growth. The real function of anting remains a mystery; other ideas neither supported nor discredited as yet include the removal of debris from feathers and skin, preparing the feathers for molting, or—suggested in a 1957 article that has understandably sunk into oblivion—autoerotic stimulation.

GOOD, AND NOT SO GOOD, GROOMING

While not as spectacular as covering one's body in ants or millipedes, another common behavior may have evolved with the main function of keeping parasites at bay, and it may be of more importance than scientists had previously realized. Most mammals and birds, and a fair share of other organisms, groom their fur, feathers, or skin. For many years, this activity seemed of only mild interest, and grooming among primates in particular was only a focus for scientists because of its social implications—the original "I'll scratch your back and you scratch mine." Partly because the importance of parasites in the lives of animals was overlooked for many years, and partly because, let's face it, a dog licking itself or a parrot rummaging through its feathers with its bill is not exactly glamorous, the more pragmatic function of ridding the body of ectoparasites was downplayed.

Anecdotes about birds with damaged bills being overrun with lice had appeared in the literature for some time, as had information about the buildup of lice on commercial chickens whose bills had been clipped to prevent the birds pecking each other. These were viewed as unusual or aberrant cases, however, and it was not until Dale Clayton, a biologist at the University of Utah, began to do systematic experiments using pigeons that the true import of preening became evident.

Clayton caught pigeons from parks and streets and fitted them with metal bits that did not impair feeding or other behaviors besides preening. After a few weeks, the birds had two to three times more feather-eating lice than birds without the headgear. Another study showed that such lice can chew away up to 30 percent of the feathers on a pigeon, which in turn makes the feathers less effective as insulation. Birds with more lice often weigh less than those with fewer external parasites, probably because having less insulation increases the metabolic rate of the host.

Mammals also spend an amazing amount of time grooming their own or others' fur, sometimes to the point that this self-care can interfere with the other demands of life, like the metrosexual whose investment in hair-care products cuts into paying the electric bill. Impala, African antelope that inhabit a savannah-woodland habitat, groom themselves frequently. In the breeding season, however, when some males defend territories and attempt to mate with as many females as possible, grooming drops sharply, and their tick loads rise far higher than those of females or nonterritorial males. Michael Mooring and Ben Hart from the University of California at Davis proposed that the impala simply don't have time to keep the parasites at bay while they are also fighting rivals, and pay the consequences. A comparison of sixty species of deer and antelope from all over the world showed that this kind of sex difference in grooming behavior was more common in species with a large disparity in male and female body sizes; such a disparity indicates that evolution has acted on males to make them more competitive and thus they would have a greater need to balance sexual activity against antiparasite behavior.

In a more prosaic example that may nevertheless hit closer to home, nine cats from a flea-infested home were fitted with those Elizabethan collars that prevent the animal from disturbing bandages or other treatments. The collars also keep the cats from grooming, and after several weeks, the collared cats had twice as many fleas as nine other cats in the same household that had been left unencumbered. (In case you are wondering, the study authors report that all of the subjects came from a "flea-infested private home housing over 30 free-ranging adult cats.")

Many animals also groom assiduously following sexual activity, and Ben Hart, who has pioneered much of the research in the area of behavioral defenses against parasites, examined the function of this behavior in rats. After introducing bacteria into the vaginas of female rats and then preventing the males they mated with from licking themselves after sex, he discovered that the bacteria were much more likely to be transferred and persist in a sexual partner if that partner could not use saliva to wash the genital region. Although saliva is not a completely effective antibacterial agent, it is still helpful in controlling infection, which suggests that concerns about keeping pets or people from licking or sucking on a wound may be misplaced.

FLOCKS, HERDS, AND SOCIAL DISEASE PREVENTION

In addition to what you do for yourself, what others do for you can be critical in reducing your chances of getting a disease.

Sheep form herds, fish school, lions are in prides, and for the esoteric there are murders of crows, bevies of roebucks, and wakes
of buzzards. Although people tend to think that all animals are
social, this is not the case. Most species are solitary, with individuals meeting only to reproduce, and being in a social group is
costly, giving rise to increased competition for food and other resources. Biologists have conjectured that those costs can be offset
if group living helps animals do one of two things: find food or
avoid predators. If, for example, you eat large animals that can be
brought down only by several hunters, as is the case for lions, you
are better off in a group. It is easier to avoid being eaten, on the
other hand, when many eyes and ears share lookout duty, as in
meerkats, small African mammals that take turns sitting upright
and scanning for predators while others feed, take care of the
young, or sleep.

Disease has traditionally been viewed as a cost of being in a
social group, not a benefit, and indeed many pathogens rely on
the close proximity of social hosts for transmission. But once the
herd or flock is formed, it may be able to provide protection from
parasites as well, and some behaviors may be helpful in keeping
parasites at bay but not at deterring predators or vice versa. For
example, many kinds of cattle form clusters with their heads low-
ered and facing toward the center of the group; this behavior
keeps face flies, a troublesome biting parasite, away from the
circle, but would obviously do little to help the inward-gazing
animals see a prospective predator. When cattle were sprayed
with fly repellant, they spent less time in bunches than untreated
cattle. Horses often stand head-to-tail, swishing flies that alight
on the partner. This is likely to be more than a minor favor, since
estimates suggest that a horse in New York State could be bitten
by about 4,000 bloodsucking flies, and lose nearly half a quart of
blood, in a single day. Howler monkeys similarly expend a great

deal of effort in combating flies, with one study showing that each monkey slaps at or otherwise avoids biting insects more than 1,500 times a day. North American caribou also form groups in the summer, when the mosquitoes of the far North are at their bloodthirsty peak. Larger groups of animals may benefit because of a kind of safety in numbers—if a roughly similar number of biting flies find a group as would find a single individual, but grow sated on just one or two hosts, each animal in the herd has a smaller chance of being selected as the victim.

Taking this idea a step farther, W. J. Freeland, a biologist working in Australia, suggested that the social organization of primates like monkeys and apes, and hence perhaps early humans, is an adaptation to avoid disease. For example, many primates make newcomers to the group go through an extended period of probation before they are accepted. During this time, the new member is threatened, bullied, and generally stressed, a situation that Freeland suggests will be likely to reveal any latent infections while serving as a kind of quarantine before the trial period ends. Many primate troops defend their territories with long-distance cries and whoops, rather than via physical contact, and while this obviously keeps individuals from injuring each other, it may also serve to lower disease transmission.

Humans certainly show a distrust of newcomers or those who appear different from their usual social group, and several scientists have suggested that xenophobia evolved as an aid to avoiding disease. A group of Canadian researchers investigated this idea using questionnaires that measured what they termed "perceived vulnerability to disease." Scoring high on this quiz means that you are less likely to agree with a statement like "I'm comfortable sharing a water bottle with a friend" and more likely to agree with such comments as "I have a history of susceptibility to infectious disease." Then the subjects, college students at the

University of British Columbia, were asked to respond to scenarios that described situations in which shunning of immigrants might be an issue, such as the arrival of a hypothetical Central African tribe to their city. As predicted, those scoring higher on their perceived vulnerability were more likely to oppose the entry of potentially germ-ridden immigrants, and potential newcomers from Eastern Europe or Eastern Asia were viewed more favorably than those from east Africa.

Two additional studies by the same group manipulated people's sensation of their vulnerability to infection by showing them a slide show either on disease or on accidents, and then assessed their attitude toward foreigners. Those seeing images of hair surrounded by bacteria or the evil that lurks in kitchen sponges were less likely to favor immigration by Nigerians than Scots, while subjects who had viewed a picture of a woman in a bathtub surrounded by electrical appliances (titled "Electricity and Water Don't Mix") showed no such bias in their views (one cannot help wondering whom they did shun as immigrants, or whether they simply developed a phobia about baths). The disease-related show subjects were also more likely to suggest that funds should be preferentially allocated to familiar immigrants rather than ones that seem more foreign.

While these results are interesting, I remain skeptical that they truly indicate an inborn or evolutionary origin for xenophobia. Certainly, shunning unfamiliar individuals keeps them from infecting you with plague, but it also has a myriad of other functions, and it is difficult, to say the least, to rule out causes other than disease avoidance for being suspicious of strangers.

Val Curtis and colleagues from the London School of Hygiene and Tropical Medicine suggested that the disgust that we feel at the sight of blood, pus, feces, and vomit, among other substances, evolved as a means to protect ourselves from becoming infected. They also employed a survey, but theirs took advantage of the BBC Web site, and they asked subjects to rate photographs of pairs of stimuli that included a plate of liquid that looked "like bodily fluids" versus a plate of blue dye, a person made up to look "feverish and spotty-faced" compared with the same person sans cosmetics, and a towel stained blue or one stained with what looked like blood. The 40,000 respondents answered questions about their nationality and gender, and then scored the pairs for which member they thought was more disgusting.

Perhaps not surprisingly, people generally tended to be more disgusted by images that one could associate with disease, and women showed a higher degree of disgust than men, although in both sexes older individuals were less likely to be disgusted. The survey also asked respondents to choose the individual with whom they would least like to share a toothbrush. Here the answers were more unexpected, at least to me; although best friends and spouses were virtually equally acceptable, the postman was least preferred, followed in increasing order of acceptability by one's boss, the weatherman, and a sibling. The authors interpret this as supporting their contention that strangers are less preferred and more disgusting because they are a larger disease threat. I can buy this for the contrast between spouse and boss, but some of the other rankings are more enigmatic. Why in the world are people so comfortable with the weatherman, whom they presumably only see on television, while shunning the poor mail deliverer? Do we really think the mailman is more likely to give us a disease than our boss? I suppose the weatherman probably has his or her own set of television lackeys who can supply vitamins, bottled water, and other accoutrements of wellness, but this seems unrelated to the evolution of disgust.

Besides, people know perfectly well that they can get sick from touching blood, pus, or vomit. This association does not show that the emotion of disgust evolved to keep us from contact with such substances. It would indeed be interesting to examine a group of people who had no familiarity with the disgusting items and see if they had an inherent response of revulsion when they encountered the material, but it would be difficult, not to say impossible, to obtain such a sample.

More convincing is a study of changes in women's perception of what is disgusting during pregnancy. Immunity is low during the first trimester of pregnancy, as a woman's body grows to accept the fetus as part of her self rather than a genetically dissimilar invader, and so one would expect vigilance against potentially infectious agents to be heightened at that time. Dan Fessler and his colleagues at the University of California, Los Angeles, gave pregnant women a questionnaire asking them about their reactions to potentially disgusting situations, like being about to drink spoiled milk. Even after taking the women's nausea from morning sickness into account, disgust levels were highest during the first trimester.

Of course, both people and animals take amazing pains to avoid feces and areas that may be contaminated with them, and while it is easy to project disgust onto a cat that fastidiously uses a litter box and would never sully another area of the house, we really do not need to attribute a particular emotion to the activity. It seems safer and more reasonable to me to note that avoidance of potentially disease-causing material is widespread among animals, just like the other behaviors I have mentioned. Calves do not graze near fresh cow dung, and other domestic animals often do not feed where they defecate. This tendency is greater in animals such as carnivores that are more likely to spread worms consumed in their previously eaten prey. Movement patterns of other animals such as monkeys are also in keeping with avoiding recently deposited feces, though it is difficult for obvious reasons to perform careful studies of the toilet hygiene of wild animals. Of course, despite the best efforts of both people and animals, infection via fecal contamination still occurs, a testimonial to the continual efforts on the part of the pathogens to overcome the defenses of their hosts.

THE BESTIAL PHYSICIAN

All of the behaviors I have been discussing are ways that people and animals keep themselves from becoming ill, rather than actually treating disease once it strikes. But the idea that animals as well as humans can actively cure themselves of ailments has been suggested many times. One of the first places people thought they saw evidence of animals attempting to heal themselves of disease was in our closest relatives, the chimpanzees. Starting in the late 1970s, researchers documented chimps in Tanzania and Uganda using plants in ways that suggested the animals were not simply adding another food to the diet. At certain times of the year, particularly when intestinal tapeworms and roundworms are most abundant, the chimps were observed taking a leaf from a plant called *Aspilia,* folding it carefully with their tongues and swallowing it whole. The *Aspilia* leaves are rough and hairy, almost certainly difficult to swallow, but a chimp may consume up to one hundred leaves at a time, usually early in the morning or at some other time when he or she has not yet done much foraging. The leaves almost certainly don't contribute to the nutrition of the animals, since still-folded, undigested leaves are often found in the dung. Similarly rough-textured plants are used in the same way by other populations of chimpanzees, as well as by gorillas and bonobos, small chimp relatives.

Several scientists, most notably Michael Huffman of the Primate Research Institute at Kyoto University, suspected that the hairs on the leaves served as scrubbers that rasped worms off the intestinal walls and helped to rid the animal of its parasites. Interestingly, although *Aspilia* also contains a powerful compound, thiarubrine-A, that could kill worms directly if it were active in the gut, the way the leaves are consumed whole almost certainly rules out their functioning as a chemical deworming agent. Swallowing the leaves before the digestive tract contains much other food presumably helps the rough surfaces contact more worms. The leaves are often swallowed by chimps with symptoms such as diarrhea, though apparently healthy individuals sometimes do the same thing. Observers have noticed similar consumption of vegetation followed by the purging of worms in geese as well as in bears before hibernation, so that the latter enter their long dormancy period without the drain of parasites on their inner resources.

Although this certainly sounds suggestive of self-medication, nailing it down with hard evidence was a challenge. Huffman and a colleague followed a group of wild chimps day after day from November 23 until February 25 looking specifically for leaf-swallowing, and were able to document just fourteen cases in twelve individuals. This does not mean the behavior is unimportant, of course; you could watch an apartment building full of people for the same amount of time and see a similar frequency of, say, aspirin use. But it means it can be difficult to come up with real demonstrations that the behavior is deliberate and occurs often enough to be meaningful in the lives of the animals. In a group of captive, parasite-free chimps without any experience of *Aspilia,* some that were offered the leaves rejected them, some ate them as if they were food, while others folded and swallowed them just as their counterparts in the wild do. Eventually, several more chimps performed the leaf-swallowing behavior, suggesting that it is possible for animals to learn it by observing others.

The other frequently observed behavior said to be a form of

self-medication by chimpanzees is called bitter pith–chewing. As the name implies, the animal takes a young shoot of a plant called *Vernonia amygdalina,* also used by local people as a treatment for diseases ranging from fever to dysentery and stomach ailments, and peels off the bark and leaves. The chimp then chews on the very bitter stem, but doesn't swallow the plant itself. Chimpanzees only seem to engage in this behavior when they are showing signs of illness like lethargy or diarrhea, and in the occasions when scientists have been able to track individuals after they chew the pith, the chimps seemed to recover within twenty-four hours, with a precipitous drop in the number of worm eggs in the feces, whereas chimps that hadn't used the plants still harbored as many or more parasites as before. The chimps also seem to recognize that the pith is not desirable simply as a food source, and mothers have slapped away the inquisitive paws of youngsters reaching for discarded *Vernonia* shoots. Pharmaceutical analysis of the plant shows that it contains several active compounds that could at least in theory help to control pathogens, especially by killing worms and their eggs.

Chimpanzees, gorillas, and other primates also consume a variety of other plants that have pharmaceutical properties, with Huffman estimating that 22 percent of the 172 plant species consumed by chimps are also used by humans to treat gastrointestinal problems or parasite infections. He also points out that the distinction between an item eaten for its nutrition and one used to maintain health or cure disease is not absolute, quoting a Japanese proverb, *"ishoku dougen,"* or "medicine and food are of the same origin." If an animal's food source also contains a chemical that can, say, kill certain kinds of microorganisms, is that animal simply benefiting from a by-product of its diet, or is something else at work? Indeed, many animals are said to use plants in a medicinal way, including elephants that supposedly induce labor

stupidity. Dogs, he grumbled, get lost every day, but "let one find his way from Brooklyn to Yonkers and the fact immediately becomes a circulating anecdote. Thousands of cats on thousands of occasions sit helplessly yowling, and no one takes thought of it or writes to his friend, the professor; but let one cat claw at the knob of a door supposedly as a signal to be let out, and straightaway this cat becomes the representative of the cat-mind in all the books." Similarly, no one remarks on the times they saw an animal suffering from injury or illness without any homemade treatment, or using a concoction that did not, in fact, do any good. Along the same lines, we tend to idealize or exaggerate the efficacy of medicines used by traditional peoples, forgetting that many old-fashioned cures were either useless or worse than the disease.

People have a soft spot for what Robert Sapolsky, a neuroscientist at Stanford University who is an expert on stress and disease in primates, called animaux savants, an idea he dismisses as "soggy with romanticism." His wariness is justified; it is all too easy to rhapsodize about animal wisdom, as well as the Noble Savage, with no basis in fact. If one wants to maintain that the animals are medicating themselves deliberately, the bar has to be set pretty high—the animal has to link feeling ill with seeking out the appropriate substance, even though the reward in terms of recovery is far removed from the action that caused it. Learning an association when the consequence of an act is delayed is notoriously difficult, even for humans. Witness for example the ineffectiveness of warning teenagers that suntans or cigarettes may cause cancer in decades to come. How could a chimp or a gorilla ever hit on the idea of peeling the bark from a twig, sucking the nasty-tasting juice, and waiting for its stomachache to go away?

Skepticism about the mechanism behind animal self-medication is behind the reluctance of some scientists to accept that nonhumans could have such elaborate behavioral techniques for fighting pathogens. The supposed self-medicating behaviors are just coincidence, according to this point of view. It would be easy to lump *Aspilia*-swallowing chimps with snipe making casts for their legs, as just another cute but improbable anecdote. But it would be a mistake to do so, not because the mechanism is poorly understood, but because it is beside the point. If animals and their parasites have been inextricably involved in a close relationship since life began, it would be unlikely for measures to deter those parasites to have evolved only in animals like apes that have the cognitive power to learn that an herb eaten today relieves pain by next week. How an animal might develop the self-medicating behavior during its lifetime is a fascinating question, but one that is separate from the observation of the behavior itself. Complex behaviors evolve the same way that other complex traits do—a little at a time, with each incremental advantage adding to the likelihood that its bearer will outreproduce those without the alteration. Bees construct honeycombs without understanding the physics of how hexagonal wax cells can be more efficiently packed than any other shape, and no one seems troubled by the accomplishment.

A recent study of the effects of malaria on mice shows that animals might naturally behave in ways that natural selection could shape into self-medication. If malaria-infected individuals were allowed to drink water with chloroquine, a quinine-containing compound used to treat the disease, their infections diminished. Interestingly, however, when given a choice between plain water and the treated stuff, the infected mice were no more likely to prefer the bitter-tasting water than were the uninfected mice in a control group. The authors of the study suggest that mammals may simply have a tendency to sample bitter foods, which could eventually lead to the kind of sophisticated behaviors seen in the chimps if the circumstances favored their evolution.

Even more intriguing is work on sheep, animals not generally known for their perspicacity. Lambs were given foods that contained mild toxins, like oxalic acid, and then given the opportunity to eat something that ameliorated the bad effects, as if they were taking medication to relieve their symptoms. Other lambs were offered the foods and curatives as well, but at different times, so that the medication didn't treat any illness. The first group learned to prefer the compound that helped their symptoms, but the second group didn't change eating habits, suggesting that it isn't so far-fetched to imagine wild animals figuring out how to self-medicate.

Even caterpillars can alter their behavior to enhance their likelihood of surviving attack by parasites. In the hills of coastal California, a kind of parasitic fly deposits its larvae into the caterpillars of a moth. The flies grow inside the body of the caterpillar, feeding on its tissue and eventually emerging, *Alien*-like, from its body cavity. The caterpillars eat either lupine or hemlock leaves, and if they have been attacked by the fly, they are more likely to survive if they consume the relatively toxic hemlock. If they are not parasitized, they do better when they feed on lupine. Given a choice, parasitized caterpillars prefer to eat the hemlock, which increases their survival rates.

This does not mean that animals—or people—are always successful at curing their illnesses, or that more "natural" cures, whatever they might be, are always best. Evolution can be a hit-or-miss affair, and all the while that natural selection is acting to make us better at evading disease, it is acting too to allow the disease organisms to overcome our defenses. Remember that our relationship with disease is a two-way street. A cure that doesn't seem to work may just be a temporary bump in the road.

CHAPTER 10

BAD, BUT NOT WEIRD:
THE REAL EMERGING DISEASES

Nowhere is the hyperbole of apocalyptic warfare analogies with disease more apparent than in news about the so-called "emerging diseases": Ebola, SARS, mad cow disease, West Nile virus, avian flu. Past decades saw Lyme disease and Legionnaires' disease, as well as HIV and AIDS. In books we are *Tracking the New Killer Plagues,* finding the *Demon in the Freezer,* entering *The Hot Zone,* and discovering *Living Terrors* amidst *The Biology of Doom.* We are exhorted about the need to marshal forces, take up arms, avert the attackers. An article in Salon.com about viruses as the "ultimate horror" in media gushes, "Uniquely fearsome, the virus goes beyond nuclear anxiety to the heart of paranoia—provoking ancient fears of disease, dehumanization, vampirism and biblical vengeance and inciting futuristic fears of human extinction." And here I thought we were just worried about dying.

Although there is no lack of headlines alerting us to approaching pandemics, both in developing nations and the industrialized world, surefire ways to avert the crisis are in short supply. Should we stockpile antibiotics? Stop eating beef? Develop more vaccines? Will the next flu season be like the devastating one in 1918–1920?

Purple prose notwithstanding, avoiding panic in the face of a disease like Ebola that causes victims to bleed uncontrollably from all body orifices is not easy. And while not all are as dramatic, none of the emerging diseases is a picnic, and for many there is no cure, no preventive vaccine, and only limited treatment. New diseases are unsettling. We used to think that even if a disease was deadly, we knew what we were up against, what the cause was. No one wanted to believe that a brand-new agent of illness could come leaping out of the air-conditioning vents, as it did in 1976 at the Bellevue Stratford Hotel in Philadelphia, where members of American Legion Post 239 caught a fatal kind of pneumonia from bacteria newly mutated to live in ventilation systems. But indeed, diseases can seemingly come out of nowhere.

Emerging diseases are not good, but they are not weird, or unexpected, or likely to be conquered—to borrow another battle-field metaphor—any time soon, or probably ever. The twentieth century brought new scourges, but so did previous ones; the advent of agriculture brought a virtual cornucopia of infections along with it. Because of this unending supply of new ailments, it is unrealistic at best and a futile waste of resources at worst to ever imagine that we can treat each new outbreak, in farthest Africa or our own backyards, as an emergency, a blip in the system, a fire to be put out with a heroic but temporary effort, and that we can get back to "normal" afterwards.

Scientists accept that new diseases will always appear. But what if there turn out to be even more infectious diseases than we had thought, more pathogens that can cause us harm? What if conditions we had always laid at the feet of inefficient Mother Nature, like hardening of the arteries or cancer or even mental disorders like schizophrenia, are caused by pathogens? These may be the real emerging diseases.

BUBONIC PLAGUE AND OTHER CLASSICS

People have known about emerging diseases and the pandemics they can cause for a long time, of course; although AIDS grabs headlines and undoubtedly has devastated many parts of the world, it still has not outstripped the Black Death outbreak of bubonic plague in fourteenth-century Europe or the 1918 influenza pandemic in terms of total mortality. Nonetheless, the idea of plagues as modern possibilities still seems shocking to many of us. The official definition of an emerging infection is one that has appeared for the first time in a population, or one that has suddenly increased in the number of people affected, or in the places where it occurs. These criteria fit many diseases at some point in their history: syphilis, smallpox, measles. Why do we believe that modern diseases are so special? Perhaps we react so urgently to twenty-first-century diseases like AIDS or SARS because some of us believed United States Surgeon General William H. Stewart, who declared in 1967 that "the war against infectious diseases has been won." Many voiced their skepticism even then, but the confident optimism can still seem warranted given the advances of vaccines, antibiotics, and other biotech tools that made the terror of the Black Death seem impossibly far away.

Or maybe we are outraged that in this day and age, when newly discovered genes are associated with everything from obesity to insomnia, we are still forced to contemplate the power of boring old germs. Germs seem so last century. An article in the *New York Times* says that we are "mesmerized by our own adorable DNA. Just last decade, after forty years of intense flirtation, this relationship was consummated as we cloned the entire human genome. Promises of improved health and longevity soon followed, as we had apparently found our way to the bedrock truths that underlie all illness." The author, physician Kent Sepkowitz, was making the point that we use faulty genes as a way to get out of personal responsibility for our healthy—or not so healthy—behavior, but I think our fascination with modern genetics also makes us less enthused about those same old, same old medical techniques of vaccination and quarantine, or of the discovery of yet another bacteria or virus.

David Morens, Gregory Folkers, and Anthony Fauci, all at the National Institute of Allergy and Infectious Diseases, a branch of the National Institutes of Health, point out that emerging infections come in two main categories, with different responses or preventive measures in each case. (They classify diseases used in bioterrorism, called "deliberately emerging" diseases, as well, but these are, thankfully, mainly hypothetical at this point.) First are the diseases that were discovered for the first time in recent decades; about thirty of these have been identified. AIDS is a classic example. Accounts of its origin differ, but small numbers of cases of HIV infection likely were present well before the disease was recognized in the 1980s. What catapulted AIDS into a global epidemic was a concatenation of events, including the movement of large numbers of people from rural to urban areas, with subsequent increases in prostitution and number of sexual partners. There are probably many other diseases that have the potential to emerge like AIDS, some with equal virulence, some with less. The key question is not which diseases are out there, buried in the jungle or seething in the swamp, but what characteristics, theirs and ours, can make them surge to our attention.

Many of the most talked-about diseases in this category, like SARS, turn out not to have much punch when they are closely examined. Robin Weiss from the Division of Infection and Immunity at University College London and Anthony McMichael of the Australian National University in Canberra devised a kind of "Richter" scale for evaluating what they term "natural weapons of mass destruction," comparing death rates due to various diseases as well as those from sources like tobacco use and traffic accidents. SARS, Ebola, and the human neurological ailment caused by eating beef infected with mad cow disease, a variant of Creutzfeldt-Jacob disease, all score remarkably low, the equivalent of a barely felt rumble that doesn't get most people out of bed. HIV, on the other hand, is a major temblor, and malaria still causes more deaths than any of the other newer, trendier infections except for those associated with HIV. Suicide and traffic accidents are in turn far deadlier than any of the headline-grabbing ailments. Few if any experts expect the truly new diseases to change their position on the scale, not because the diseases themselves are not virulent—remember that bleeding-from-all-body-orifices aspect of Ebola—but because we are starting to get a better feel for their attributes. SARS was controlled because an all-out effort by world health authorities kept it from spreading from person to person, not because it turned out to be a benign ailment after all. Weiss and McMichael didn't include bird flu in their analysis, but similar vigilance may well spare us again. Other diseases might turn out to be a larger threat, but not because they are novel. Instead, the biggest worries are those that have been with us, in one form or another, all along.

THEY'RE BAAACK...

The second kind of emerging disease is shaping up to be more of a public health problem than the truly new pathogens, despite the latter having more shock value in the media. These are the reemerging or resurging infections, those diseases that may have been major killers in the past but subsided, only to regain our attention in modern times. Unquestionably we have done very well at reducing infectious diseases worldwide; deaths from infections have dropped dramatically since 1900, with smallpox eradicated completely, and polio nearly so. But other diseases, like tuberculosis, have gone from nearly being vanquished to being of renewed concern. Death rates from tuberculosis actually increased in the United States in 2002, and the foreign-born are more than eight times more likely to suffer from the disease than natives. Tuberculosis, and the bacteria that causes it, are, theoretically at least, reasonably easy to control. Even if exposed to the bacteria, a healthy person can often keep from becoming ill and spreading the infection. But infection with HIV suppresses the body's immune response. Thus when people are HIV-positive, even small numbers of tuberculosis bacteria, or an infection that was previously suppressed, can begin to multiply and progress to a contagious form of the disease. Widespread poverty, with its accompanying lack of hygiene and close quarters, also fosters tuberculosis, and drug resistance, as I discuss next, compounds the problem.

Several other diseases, such as Lyme disease or monkeypox, are not new, but used to be rare, causing little illness among the global population. These diseases were unfamiliar to Westerners until modern circumstances caused their expansion, with Lyme disease brought to our attention because suburban expansion meant more people came into contact with the deer and ticks that bore the infection. Monkeypox, which ordinarily infects a variety of African rodents in relatively benign obscurity, arrived in the United States along with exotic pets like the giant Gambian rat. Rodent fanciers liked the plus-sized animals because, as the head of the Rat and Mouse Club of America said in a 2003 USNews.com article, "It's kind of hard to really hug a domestic rat." Hug me, hug my parasites.

The most worrisome reemerging diseases, however, are entirely of our own manufacture, and I am not referring to bioterrorism. Antibiotics revolutionized medicine when they became widely available after World War II; bacterial diseases that had been difficult or impossible to cure, like gonorrhea, syphilis, tuberculosis, and some forms of pneumonia, suddenly seemed on the verge of disappearing. But physicians soon began to notice that some antibiotics were no longer effective. These days, antibiotic-resistant bacteria are such a serious public health concern that the World Health Organization published this anonymous bit of doggerel titled "The History of Medicine":

2000 B.C.	Here, eat this root.
1000 A.D.	That root is heathen. Here, say this prayer.
1850 A.D.	That prayer is superstition. Here, drink this potion.
1920 A.D.	That potion is snake oil. Here, swallow this pill.
1945 A.D.	That pill is ineffective. Here, take this penicillin.
1955 A.D.	Oops...bugs mutated. Here, take this tetracycline.
1960–1999	39 more "oops." Here, take this more powerful antibiotic.
2000 A.D.	The bugs have won! Here, eat this root.

Antibiotics like penicillin, tetracycline, and ampicillin are bacteria killers of the finest kind. They are such effective drugs because they target bacterial cells and leave the rest of the cells of a vertebrate body alone. They attack the wall of the bacterial cell, a structure we and other complex multicellular creatures lack. Viral diseases like colds are impervious to the effects of antibiotics because viruses invade the cells of their host and do not have a cellular structure of their own. Scientists have always known about this limitation, but because the results of antibiotic therapy can be so dramatic, people almost immediately began demanding antibiotics for all kinds of illnesses. Making matters worse, patients sometimes use the drugs carelessly, without completing the full regimen of doses.

The reason these deviations are a problem lies in the same simple fact that is the focus of this book: If you do something to bacteria, bacteria will do something back to you. Antibiotic resistance is nothing more than evolution by natural selection, like beak size changing in Darwin's finches in the Galápagos, but far more deadly.

Imagine a population of bacteria within a human host. There are thousands of individual bacterial cells, each with its own genetic makeup. Although many of the bacteria are genetically identical, having arisen through asexual reproduction, some have exchanged genes with other bacterial cells in a process called, in that racy way scientists have, conjugation. Other genetic material can be transferred through a variety of peculiar processes found only in bacteria. Just as populations of humans will vary in traits large and small, the bacteria vary as well, and a few of them, through chance mutation alone, may happen to be able to resist the effects of antibiotics a bit longer than their neighbors. If an antibiotic is given for too short a time, or for the wrong reason, those variant bacteria will linger unscathed, and then multiply in the body. Administering another dose of the antibiotic after this occurs will be futile, because the mutant forms will now dominate the bacteria population. It isn't that the bacteria change in response to the antibiotic, like some evil conscious creature able to detect the onslaught of powerful drugs, it's that there are always a few lucky individuals around to take advantage of a changing environment. Antibiotics are so widely used, and bacteria can spread so easily, that resistant strains rapidly overtake susceptible ones in an entire population. In an article on the biological costs of anti-biotic resistance, microbiologists Dan Andersson and Bruce Levin bluntly stated, "The use of antibiotics by humans can be seen as an evolutionary experiment of enormous magnitude."

The consequences of that experiment have been equally enormous. Nearly a third of infections with *Streptococcus pneumoniae,* the bacteria that causes a form of pneumonia, meningitis, and ear infections, are resistant to penicillin. Tuberculosis, syphilis, typhoid, gonorrhea—all have antibiotic-resistant strains in some parts of the world. In southeast Asia, 98 percent of all gonorrhea cases are resistant not just to one but to multiple anti-biotics, which enhances HIV transmission. *Shigella,* the bacteria that causes a common type of dysentery, has become resistant to virtually every drug available, and it is now back to killing people worldwide. Because of the frequent use of antibiotics in hospitals, infections acquired there are particularly problematic; more than 70 percent of the bacteria causing infections in people while they are patients in hospitals are resistant to at least one of the drugs commonly used to fight them. Many bacteria are resistant to multiple drugs, and usually a drug used as a second, third, or last resort is far more expensive and difficult to obtain and use than the more common treatment. This need to go through numerous drugs also prolongs the time the patient suffers from the infection.

A major source of antibiotics in our environment is domestic animals. Given routinely to poultry and cattle to encourage growth and treat infections, more than 29 million pounds of anti-biotics are used each year in the United States, according to the Union of Concerned Scientists. The more conservative estimate by the Animal Health Institute is 18 million pounds, still hardly a paltry amount. Here the concern is that selection for resistant bacteria in animals will allow those pathogens that can infect both animals and people to become resistant to antibiotics used to treat human diseases. In Europe, several antibiotics have already been banned in agricultural use, and early results indicate that antibiotic resistance is on the wane.

The rest of the world has been less successful. Bruce Levin, writing this time with Dutch medical researcher Marc Bonten, points out that despite the call for more prudent use of antibiotics, "The only slowing that has taken place is the pace at which antibiotics with new targets are entering the market." In other words, the bacteria are winning.

The reasons are numerous and complex. Farmers are pressured to produce meat, eggs, and milk as quickly as possible, and they use antibiotics to prevent disease, to keep any illnesses that do arise from spreading, and to tamp down the digestive-tract bacteria that are a natural consequence of the grain-based diet many domestic animals are fed. Patients want antibiotics for everything, and physicians sometimes accommodate their wishes, despite antibiotics being useless against viral diseases, including most childhood ear infections and the sinus inflammation and pain that sometimes follows a cold. (Even the conventional wisdom that local facial pain, green or yellow nasal discharge, or discharge only on one side indicates a bacterial infection turns out to be untrue.) Dr. Ralph Gonzales at the University of California, San Francisco, lays the blame squarely at the feet of educated patients—in a 2005 *New York Times* article, he refers to them as "30- or 40-year-old professionals with bad colds and overwhelming deadlines"—who wheedle their physicians into giving them treatments they don't need.

In some countries, counterfeit drugs containing only a trace of the actual antibiotic abound, and folklore contributes to misuse of the drugs; in the Philippines, a drug used to treat tuberculosis is believed to be a "vitamin for the lungs" and hence given freely to children. Pharmacists in many developing nations lack training, and the cost of even common antibiotics like penicillin can be prohibitive, encouraging undertreatment and the sharing of drugs among patients and their families. World travel has quickly taken the problem to far reaches of the globe. Among the isolated Wayampis Amerindians of French Guyana, antibiotic resistance was detected even in those individuals who were not taking any drugs and who had not been hospitalized.

The economic costs of drug resistance—longer hospital stays, use of more expensive alternative medications, and higher death rates—have been estimated at anywhere from $150 million to $30 billion a year, depending on exactly how you crunch the numbers. The medical establishment is well aware of the problem, and organizations including the World Health Organization, the Centers for Disease Control and Prevention, and the National Institutes of Health are all attempting to deal with it. A recently developed vaccine against *Streptococcus pneumoniae* shows promise for circumventing the need for treatment with antibiotics or anything else.

People sometimes fear that forgoing antibiotics and antibacterial soaps means leaving ourselves vulnerable to an onslaught of disease-carrying microbes. But products like simple soap or alcohol-based cleansers can kill bacteria without containing antibiotics, and without causing problems like the evolution of resistance. Unlike antibiotics, which target specific elements of the bacterial structure, soap binds indiscriminately to the bacterial cells and allows them to be washed away. It is easy for bacteria to evolve resistance to a substance with a narrow objective, much harder to evolve resistance to a generalized foe. And soap and water, or bleach and ammonia-based cleansers, do not linger on surfaces the way that antibiotics or antibacterial cleansers do, which means there is less danger of killing off innocuous bacteria as well.

WHAT DID WE DO TO DESERVE THIS?

Although people tend to imagine Africa and other parts of the tropics as the steamy source of new and frightening pathogens, a survey of disease outbreaks from 1940 to 2004 by Peter Daszak of the Consortium for Conservation Medicine shows that most emerging diseases currently attracting attention originated in Europe, North America, and Japan. Ebola and Marburg get a lot of press, but they seem to be exceptions. The tropics do contain a lot of diseases, just like they contain a lot of kinds of frogs, flowers, and flies, but a French group of scientists led by Vanina Guernier found that the diversity of disease-causing organisms followed the same rules that the larger species did, with more different types of living things closer to the equator. The tropics may simply form a bigger source pool for disease, but diseases do not emerge from them disproportionately.

Diseases do emerge—or reemerge—because any time a plant, an animal, or a person goes somewhere, its diseases are likely to go with it. Global travel means that an infection can be transported to a rich new source of hosts within hours. And if, once it moves, it lives in crowded conditions near new kinds of neighbors, then the stage is set for diseases to jump rapidly from individual to individual, even across species. Crowding, particularly when accompanied by poor sanitation, provides an incubator for diseases like cholera, which is spread in water contaminated by sewage. The nineteenth-century improvements in public sanitation "probably saved more lives than all the twentieth-century vaccines and antibiotics together," according to Weiss and McMichael, but these advances are not permanent, and events like the 2004 southeast Asian tsunami or the 2005 earthquake in Pakistan, as well as less catastrophic situations like urban poverty, continually threaten to undo them.

Many, perhaps most, of the new diseases in the headlines are zoonoses, diseases that have spread from animals to humans. By one estimate, 75 percent of human emerging diseases fall into this category. In some cases, such as avian flu, health officials watch anxiously for the disease to mutate so that it can then be transmitted directly between people, without the domestic poultry intermediary that is now needed. Others, like West Nile virus, are harbored in a variety of kinds of wild animals—birds, in the case of West Nile—where they remain at high enough levels that they not only pose a risk to the wildlife themselves, but can keep transferring to humans via mosquitoes that bite both the birds and people.

A classic example of a recent zoonosis is mad cow disease and its counterpart in humans, variant Creutzfeldt-Jakob disease. Mad cow, more properly called bovine spongiform encephalopathy (BSE), is caused by a kind of oddly folded protein called a prion, and was first diagnosed in the United Kingdom in 1986. It causes neurological symptoms like impaired movement, and its spread was greatly enhanced by feeding ground-up infected cows to calves. BSE reached a peak in the UK in 1993, when nearly 1,000 new cases were detected each week, and at the end of April 2005, more than 184,000 cases had been reported.

Getting sick from cannibalized brains and spinal parts is bad enough when it happens to cows, but people really became agitated when the disease was found in humans. Variant Creutzfeldt-Jakob disease also produces neurological symptoms, including dementia and other psychiatric abnormalities, and it appeared in the UK in the mid-1990s, about a decade after people were likely exposed to BSE-contaminated meat. That time frame is consistent with the expected incubation period for the "regular" form of Creutzfeldt-Jakob disease, and other laboratory evidence also points to people having caught the disease from cows. Of the 190 cases of variant Creutzfeldt-Jakob disease reported as of March 2006, all but 30 were in the UK. The current risk of the human disease and the incidence of BSE in cattle appear to be very low in the United States, though the Centers for Disease Control continue to monitor the situation. Neither BSE nor the human form of the disease is poised to explode into an epidemic, but they illustrate the ease with which we can share illnesses with animals.

Older diseases probably originated this way too, with smallpox, for example, being very closely related to cowpox in domestic cattle, but they largely arose many thousands of years ago, when agriculture and the domestication of animals meant that humans shared space, water, and microbes with a variety of species for the first time. What has changed over the last several decades is the degree to which humans are encroaching on wildlife in areas that used to be relatively pristine. Bushmeat, meat from wild animals, is a particularly rich source of diseases, especially where people hunt monkeys and other primates. The more closely related a species is to humans, the more likely that a disease will be able to jump from its usual host to a human one. People eat game both as a luxury, as in Asia where exotic species like civet cats are prized, and out of necessity, as in Africa where monkeys and apes may be the only source of protein available to people displaced from their homes and farms. Although exact figures are hard to come by, it is thought that over a billion kilograms of wild animal meat are consumed each year in Africa, and anywhere from 67 to 164 million kilograms in the Amazon basin. The problem is compounded because sick animals are easier to capture; after a single infected chimpanzee was butchered, 21 people died of Ebola through contact with the carcass. Wild animals kept as pets, like the Gambian rat mentioned above, are another source of new diseases, though we can also get sick from familiar animals like cats and dogs.

Even sticking to domesticated animals for meat rather than hunting wild ones is not a guarantee against disease; in Malaysia, the Nipah virus killed 105 people in 1998. The virus also occurs in pigs, and the people contracted it on pig farms, probably from the same kind of contact that gives poultry farmers bird flu, but the pigs themselves seem to have become infected from the droppings of fruit bats feeding in the trees on the farms. Over a million pigs were culled, and farmers in other parts of Asia are watching the situation with unease. Nipah virus seems to be endemic among even healthy bats, and it may occur in other animals as well. Who knew that the man who ate the pig would also have to consider the bat that ate the fruit?

BIODIVERSITY AND PATHOGEN POLLUTION

Emerging diseases are enough of a concern because of their direct effect on the people suffering from them. But scientists are increasingly realizing the subtler but equally troubling effects of emerging infections on the environment. So-called pathogen pollution happens when parasites get to new hosts because of human intervention, sometimes accidentally, when a species like the zebra mussel floats into new ports on the hulls of ships, and sometimes deliberately, when an ostensibly helpful but foreign species like a ladybug is introduced for biological control of a pest. In either case, the invader's pathogens can have fertile new ground to colonize. And introduced pests like the mussels are bad enough in their natural state, but if only a few individuals are introduced and these happen not to be infected with disease, the introduced species can multiply even faster because its population growth is unchecked by pathogens. Even conservation efforts to release captive-bred animals from zoos back into the wild can inadvertently send foreign disease organisms with the endangered host.

People and their planes, trains, or cars don't even have to physically haul the animal and its diseases around; climate change, now acknowledged to be at least partly human-induced by all but the most skeptical, can do it for them. As animals expand their ranges in areas affected by global warming, their worms, fleas, and flukes are expanding along with them. A roundworm that infects the lungs of musk oxen, sturdy buffalo-like animals that inhabit Arctic regions, has suddenly become far more common, threatening the demise of populations in some places. It turns out, however, that the parasite is not actually new, but one that used to occur at very low and relatively harmless levels. A warmer climate means that the worms develop much more quickly, and can spread more easily and in greater numbers to new hosts. Change the species distribution and you will change the dynamics of disease.

Parasites can also remain as the ghosts of past species introductions. A lung fluke now threatening two species of Costa Rican frogs is new to Central America; it apparently came from bullfrogs, which are native in much of North America and have been introduced to many other parts of the world. The bullfrogs themselves have disappeared from Costa Rica, but their parasitic legacy lingers. Several other amphibian diseases may have come along with their froggy immigrant hosts, and are thought to be contributing to the worldwide decline of native amphibians that scientists have observed over the last several decades. In the highlands of Central America, the spectacularly colored harlequin frog, the equally flashy golden toad, and many of their relatives are at the brink of extinction or have already vanished. Their demise is attributed to a pathogenic fungus that in turn was aided by global warming, adding disease as another link in the chain of climate change's effects.

Emerging diseases can also turn back to their original host sources, with exacerbated effects if those hosts have become endangered. Ebola is deadly in several species of monkeys and apes as well as humans, and in 2003 an outbreak of the disease in gorillas, aided in its spread by humans, led to the death of 600–800 of the great apes, a substantial fraction of the world's gorilla population. This back and forth of disease between people and animals means that monitoring wildlife disease outbreaks can help alert public health officials to possible disease risk in humans. Peter Daszak from the Consortium for Conservation Medicine points out that emerging diseases of humans, wild animals, and even plants are all linked. William Karesh and Robert Cook of the Wildlife Conservation Society talk about an approach to emerging diseases that is "based on the understanding that there is only one world—and only one health." In other words, whether one is concerned about endangered species and saving the rain forest or not, the fate of wildlife is intimately involved with the fate of humans, via disease.

Dan Brooks, a parasitologist at the University of Toronto and a large, genial man given to quoting from nineteenth-century biologists like Thomas Huxley, calls emerging infectious diseases "evolutionary accidents waiting to happen." He has long been a proponent of what some would call old-fashioned taxonomy, calling for an inventory not of the glamorous species like whales and eagles, but of the parasites of the world. We need a list of every worm, every fluke, and every one-celled parasite, and we need to characterize the evolutionary relationships among them. Only then, he claims, will we be able to assess the risk of diseases passing from one animal species to another and then potentially to human beings. For instance, white-tailed deer in Costa Rica harbor eighteen kinds of parasites, including six kinds of ticks. None of the ticks in turn are species that carry the microbe that causes Lyme disease, but they are close relatives of ones that do, and so it is reasonable to expect that Lyme disease could become a problem if even a small number of infected ticks make their way to Costa Rica, because the microbe could transfer to the new tick species with relatively little difficulty. We will never be able to evaluate the risk of this and similar occurrences until we know what the possible alternative vectors for the disease might be.

The conventional wisdom among biologists proclaims that most parasites are so specialized that they pose little risk of switching hosts, but Brooks and his colleagues question this immutability, pointing to numerous exceptions. If a parasite is good at exploiting one host, another one with similar physical characteristics will be an easy mark, even if the new host is not evolutionarily similar to the old one. Thus, hypothetically at least, a parasite of a squirrel might be able to live inside of a marsupial honey glider, even though the two have not shared a common ancestor for many millions of years, simply because the two hosts have similar vulnerabilities and physiology. Dan Brooks and Amanda Ferrao examined several kinds of parasitic worms in two groups of primates, looking for species that inhabited multiple hosts, and determined that nearly a third had jumped to a new host. Human activities that place us in close proximity with wildlife, like massive refugee movements after global conflict, heighten the risk of a parasite switching to a human host.

To deal with the problem of emerging infectious diseases in wildlife, in the recognition of Karesh and Cook's "one world, one health," Daszak calls for the formation of interdisciplinary teams, with veterinarians collaborating with political scientists, physicians, public health officials, and wildlife biologists. These teams would be better equipped to monitor the different facets of a disease than individuals concerned only with a pathogen's effect on a subset of people or animals. West Nile virus, for example, is not just of concern because of its potential for infecting people; it infects so many different kinds of birds, including many of our most beloved songbirds, that it could produce a silent spring of a different sort than that feared by Rachel Carson. If the virus spreads to relatively isolated places like the Galápagos or Hawaiian Islands, entire species could be wiped out.

Even if one is indifferent to the fate of Hawaiian honeycreepers or Galápagos finches, however, pathogens of wildlife can be a major concern for humans. Do you like salmon, with all those healthy omega-3 fatty acids said to be so good for you? An Asian fish originally introduced into ponds near the River Danube—and now called "the most invasive fish species in Europe"—is already causing a European fish species closely related to salmon to die, and those that remain do not spawn when the intruder is present. The fish carries a parasite inside its cells that is benign to the Asian species but pathogenic to the European fish. All the European species had to do was share water with the Asian species to suffer from the effects of the parasite, and the authors of the study, led by Rodolphe Gozlan from the UK, suggest that the pathogen could eventually threaten wild or farmed salmon and trout around the world.

THE REAL EMERGING DISEASES?

Most people, especially in Western countries, do not die of plague or malaria or AIDS; they die of chronic ailments like cancer or heart disease, their arteries clogging over decades until finally the heart itself can no longer function. Why do we get those kinds of diseases? Until recently, most physicians and medical researchers shrugged, positing a complex combination of faulty genes, environmental factors like diet or exposure to toxins, and the inevitability of aging. No one element could be pinpointed as the cause of a disease like cancer, so efforts to cope with such diseases are multipronged, with some people trying to determine if low-fat diets help and others researching genetic links to susceptibility. The generally accepted idea is that these diseases are just the body malfunctioning for unknown reasons, like a car that develops transmission trouble.

But what if they are wrong? What if diseases like cancer or atherosclerosis, the "hardening of the arteries" that precedes some forms of heart disease, are caused by infectious agents, just like flu or bubonic plague? Paul Ewald, the biologist who brought us the idea of managing virulence and who questioned the inevitable evolution of a pathogen to benign coexistence with its host, thinks that is likely to be the case. He and others are starting to wonder if we have not been too quick to dismiss pathogens as causes for an enormous variety of ills, not just cancer and heart disease but Alzheimer's disease, multiple sclerosis, and mental illnesses like schizophrenia and obsessive-compulsive disorder.

The poster child for the microbial cause of an illness previously believed to be systemic rather than infectious is the humble stomach ulcer. Before the 1980s, the medical establishment listed a potpourri of causes for ulcers, ranging from stress to a bad diet to overuse of aspirin. People with ulcers were told to drink milk to coat their stomachs, to avoid alcohol, to reduce stress, and eat a bland diet. Laudable pieces of advice all (perhaps with the exception of the spice-free diet, given the lessons of the last chapter), but as it turned out, completely irrelevant to the illness itself. Ulcers, it was determined following some clever and dogged work by Australian physician Barry Marshall and pathologist J. R. Warren, are caused by the bacteria *Helicobacter pylori*. The bacteria are present in many of us, but in some they turn rogue and cause lesions in the lining of the digestive tract. Antibiotic therapy for a month cures 90 percent of cases. In 2005, the Australians shared the Nobel Prize for their work.

A happy ending, to be sure, except that the idea took decades to catch on, first meeting with skepticism and downright dismissal. Even now some doctors do not test patients with symptoms of ulcers for the bacteria. Medical scientists are not comfortable with reassigning blame for chronic diseases to infectious agents. Ewald believes that part of the problem is that they have learned the set of criteria for determining the microbial cause of a disease all too well, and then cling to it in the face of evidence to the contrary. Those criteria are called Koch's Postulates, named after the 1905 Nobel Prize winner Robert Koch, who established with painstaking experimentation the kind of bacteria that were responsible for anthrax. Koch's Postulates outline the steps required to demonstrate that a given microbe causes a disease. One must find the putative disease agent in each patient, then isolate and grow the pathogen by itself, use it to infect a healthy individual, and finally recover the same pathogen from the experimentally infected animal.

These stringent requirements are simply not feasible for many diseases with effects that take place months or years past infection, argues Ewald. We already know that some cancers, such as adult T cell leukemia, are caused by viruses, but they take decades to show up, even though people are infected as infants via their mother's milk. Furthermore, not all infected individuals develop the disease. Ewald believes that many more—perhaps all—cancers could have similar pathways to illness. Koch's Postulates are simply unworkable in such a situation, but that does not mean that infectious agents do not cause the leukemia and many other diseases where it is impossible to extract and culture the microbe and reinfect another healthy individual in a realistic time frame.

Ewald and his colleague Gregory Cochran also believe that the sheer magnitude of chronic diseases like cancer and heart disease argues for their infectious cause. As an evolutionary biologist, Ewald found the notion that a large proportion of a population would perpetually suffer from a disease that reduced the likelihood of reproduction to be simply implausible. The relentless sieving process that is natural selection should have eliminated hearts and other organs subject to defects like atherosclerosis many generations ago, if indeed genetic predisposition is the main reason for susceptibility to such diseases. Even small reductions in the number of offspring an individual has can result in the eventual elimination of that individual's genes. Although diseases that strike late in life will have a smaller effect than those that cut off an individual's reproduction at the start, Ewald maintains that because humans spend so much of their lives contributing to their children and grandchildren, anything that causes earlier death or disability will be costly. No such rules, of course, govern the course of infectious diseases; as long as they can spread, they will. Therefore, Ewald and Cochran argue, if diseases persist in killing off or sickening a lot of people, we should suspect that those diseases are caused by pathogens rather than deficiencies in our constitutions. It's not that our bodies should be perfect, but that selection will favor a less-sickly and higher-reproducing form over any others. Furthermore, studies of supposedly genetic diseases in identical twins, who have the same genes, rarely show that the illness occurs in both members, as would be expected if genes dictate the ailment.

No one denies that genes and environmental contributors like diet and exercise play a role in cancer and heart disease, but virtually all diseases, including unarguably infectious ones like tuberculosis or influenza, are influenced by these factors. Women and men differ in their susceptibility to flu, for instance, but the same virus causes the disease in both sexes. People who smoke are more likely to get colds, but the cold is caused by a virus, not tobacco. Certainly physicians often talk about diseases as having a "multitude of causes," but Ewald sees this as a meaningless cop-out. Just because several things affect a disease doesn't mean they all have an equal part in its cause. We have to pinpoint the real causative agent, he says, or we won't get anywhere with a cure.

Ewald tends to take the maverick role when he discusses how his ideas are dismissed by conventional medicine. Bold—and unproven—statements like "Less than 5 percent of all cancers are known to be caused without any assistance from infectious agents," from his book *Plague Time: The New Germ Theory of Disease,* are bound to raise some hackles. But conventional medicine has long recognized that pathogens play a role in at least some cancers, with Francis Peyton Rous discovering in 1911 that tumors in chickens could be transmitted virally; he won the Nobel Prize for his work on what is now called Rous sarcoma virus. More recently, the human papilloma virus has been linked to cervical cancer. The question is how widespread such infectious causes are.

A recent colloquium of the American Academy of Microbiology focused solely on potential microbial causes of chronic diseases. Candidates included the bacteria *Chlamydia pneumoniae* in atherosclerosis as well as Alzheimer's disease, Epstein-Barr virus in multiple sclerosis, enteroviruses in diabetes, and prions, the quirky little pathogens implicated in mad cow disease, as a possible cause of amyotrophic lateral sclerosis, also called Lou Gehrig's disease.

The colloquium members pointed out numerous difficulties in pinpointing an infectious cause of any of these diseases. The same set of symptoms may be caused by different microbes, so that a cytomegalovirus has also been implicated in atherosclerosis along with the *Chlamydia;* conversely, Epstein-Barr virus might be a cause of chronic fatigue syndrome as well as multiple sclerosis. At the same time, a large number of people show signs of infection with Epstein-Barr, which can cause mononucleosis, or they can also remain completely asymptomatic. Why do some people with the virus develop a disease like multiple sclerosis and others do not? The answer isn't clear; maybe the virus isn't really the cause of the disease, or maybe the time that the virus is contracted is key, with an early infection providing less risk of later complications. In addition, some of the microbes recently found to cause diseases like mad cow, called prions, are extremely difficult to detect, requiring sophisticated technology not commonly used in diagnosis. On a more mundane note, if a disease is located at a so-called "nonsterile site," meaning a place in the body that ordinarily teems with microorganisms good and bad, like the gut, the isolation and detection of a single culprit for a disorder is daunting. Even finding the pathogen in the disease site is far from being a smoking gun, since bacteria may simply congregate where damage from other sources is already done.

An even more out-there suggestion about the infectious nature of a chronic disease involves a condition only some people are willing to call a disease at all: obesity. Physicians and scientists have long acknowledged that a highly complex combination of attributes make people obese, including genes, childhood environment, activity levels, and of course diet. But a few researchers are convinced that either the kinds of microbes in a person's digestive tract—those bacteria responsible for Joel Weinstock's favorite line—or one of a handful of viruses can contribute to obesity in people who otherwise eat diets that would not make others fat. Such causes would go a long way toward explaining why a group of similarly sized people can eat the same number of calories with some losing weight, some gaining it, and others remaining stable. It's too soon to tell. But if microbes can affect so much else about us, it makes sense that they could influence how much we weigh.

THE EMERGING INFECTIOUS MIND

What about mental illnesses? Some of the most intriguing evidence for infectious causes of chronic disease comes from disorders long thought to have little or no physical cause. The idea that some mental illness has a physiological cause is not new, of course; syphilis, for example, causes insanity in its later stages, and pellagra, a vitamin B deficiency disease, used to send poverty-stricken Southerners to mental institutions before a doctor with the Public Health Service, Joseph Goldberger, recognized that their nutritionally poor diet of cornbread, molasses, and pork fat was to blame for their dementia.

Mental diseases have been primary targets of the bland "combination of factors" explanation that Ewald scoffs at. Again, no one doubts that symptoms of many conditions, from schizophrenia to bipolar disorder, ebb and flare up at different times, possibly due to stress, diet, or some other environmental trigger. But is there an underlying infection that causes the disease in the first place?

Two of the hottest areas for research into infectious causation are obsessive-compulsive disorder and schizophrenia. A form of OCD that appears relatively suddenly in children, so that they go from normal play to performing repetitive behaviors like hand-washing or book-straightening virtually overnight, has been linked to infection with streptococcus bacteria, the same microbe that causes "strep throat." This form of the disorder is called PANDAS, for pediatric autoimmune neuropsychiatric disorders associated with streptococcal infection. The theory, most vigorously supported by Susan Swedo, a pediatrician and medical researcher at the National Institutes of Mental Health, holds that the antibodies produced in response to strep infection damage the part of the nervous system implicated in OCD. Antibiotic therapy, if given early enough, should then cure the disease, or at least keep it from progressing.

Of course, things are not that simple. To qualify as PANDAS, the disease has to be associated with strep to begin with, which means it is something of a self-fulfilling prophecy. Everyone agrees that infection with bacteria will never explain all forms of OCD, even in children. Testing children with sudden-onset OCD for strep antibodies shows an association that Swedo and others claim is far higher than would be expected through chance alone, but skeptics maintain that the correlation is just that, and that no causation has been shown. Antibiotic therapy has had mixed results, and while some physicians and parents advocate taking penicillin after the first episode of PANDAS, to prevent recurrence of symptoms, this idea is also controversial. Because the antibodies to the bacteria have already formed, and presumably had a chance to do their damage, by the time PANDAS is diagnosed, just treating the strep itself is not enough. Current thinking is that PANDAS is rare, but real, explaining an as-yet-undetermined proportion of cases of OCD in children. Research is continuing, with the hope of finding a cure for even a small proportion of sufferers.

The link between infectious agents and schizophrenia is even more tantalizing. Schizophrenia is a disorder that virtually cripples its victims, who suffer from a variety of disturbances ranging from auditory or visual hallucinations to delusions to memory loss and an inability to communicate. It is often diagnosed in young adulthood, and over the years has been attributed to a variety of causes. Most recently, genetic defects have been implicated in schizophrenia, so that having a family member with the disease increases an individual's likelihood of developing it. Genetics do not explain everything about the incidence of schizophrenia, so many psychiatrists and researchers believe that the family environment and other developmental factors influence whether a given person with the defective gene will actually develop the disease.

But Ewald and Cochran, along with Levi Ledgerwood, a scientist at Massachusetts General Hospital, believe that schizophrenia is the perfect candidate to have an infectious cause, given the low reproductive rates of schizophrenics and the cluster of other disorders, including higher rates of metabolic disorders, diabetes, and thyroid disease, in schizophrenics and their parents. Evolution should have knocked out the genes associated with schizophrenia years ago, since it carries such a high cost, they say. They point to the seasonality of births of people who later develop the disease; in temperate climates, schizophrenics are more likely to have been born in late winter or early spring. Furthermore, schizophrenics tend to be born at times of year when stillbirths, often caused by prenatal infection, are also at their highest in a population. To Ewald and his colleagues, this argues for schizophrenia being an infection that the mother contracted during early pregnancy, and suggests that the culprit is something she is more likely to be in contact with when she spends time indoors.

They and others have suggested several pathogenic candidates, including a virus called HSV-2, but the strongest one to date appears to be toxoplasmosis, a very common parasite that is ordinarily transmitted from rodents like mice and rats to predators like cats. Toxoplasmosis alters the behavior of the hosts it infects in many subtle ways, but if a female mammal is infected during pregnancy, it can also damage the central nervous system of her fetus. Pregnant women are urged not to come in contact with cat feces and to avoid changing the litter box, but exposure may occur even if these precautions are taken, if, for example, cats excrete the parasite in garden soil. Children of mothers who were infected with toxoplasmosis during pregnancy show higher levels of mental retardation, brain malformations, and seizures. Ewald and coworkers believe that the reason more people with schizophrenia are born at certain times of year is that their mothers were more likely to have come into contact with cats and their feces. Experimental evidence of the link between toxoplasmosis and schizophrenia, however, has been hard to come by. Determining whether laboratory animals such as rats suffer from hallucinations if their mother had toxoplasmosis is a challenge, to say the least, and obviously humans themselves cannot be used as subjects.

So a group of scientists from New York and California, led by Alan Brown of the New York State Psychiatric Institute, did the next best thing. Using a huge sample of mothers enrolled in a child health study between 1959 and 1967, the researchers tested blood taken during pregnancy for antibodies to the *Toxoplasma* parasite, comparing 63 cases in which the babies born to the mothers had later developed schizophrenia with 123 cases of mothers whose children did not exhibit the disease. If a mother had high levels of the antibody, indicating an infection, her child was more than twice as likely to become schizophrenic as one from a mother with low levels. In a paper published in the *American Journal of Psychiatry,* Brown and his collaborators conclude that "Given that toxoplasmosis is a preventable infection, the findings, if replicated, may have implications for reducing the incidence of schizophrenia." Typically cautious language from the medical establishment, but nothing less than revolutionary compared with the theories of only a few decades earlier.

No one, including Ewald, would suggest that every case of schizophrenia or OCD is caused by a pathogen. But the data are compelling, and with the specter of bacterial-induced ulcers before them, physicians are being forced to reconsider those dull but ubiquitous parasites as the cause of more diseases than ever.

CHAPTER 11
WHO'S IN CHARGE HERE, ANYWAY?

It's a warm spring day, filled with bird song and the sound of morning traffic. You awaken with a sense of well-being. You take your morning walk, feeling surprisingly energetic. In fact, it seems as if you could keep going forever, enjoying the air and the sunshine. Eventually you get to a small park, and although you don't ordinarily like walking in the woods, somehow today it feels right, and you veer off onto a path that takes you down to a small stream. Suddenly, you begin to feel a bit peculiar, light-headed. For reasons you cannot explain, the water itself seems inviting, glinting in the sunlight as it slides over the moss-covered rocks. You glance uneasily around, a bit self-conscious but unable to avoid an overwhelming urge to splash into the stream, not stopping to remove your clothing.

The light-headed feeling intensifies, and you grow quite faint. Why had you wanted to go to the stream in the first place? You never liked swimming, and you do not even eat fish. As you sink into the water, you feel an odd tightness in your chest, and then dimly perceive to your horror that your body is literally splitting down the center. A huge writhing worm, coiled like a garden hose, emerges from your chest. You collapse back in the water, limp and bleeding, barely conscious, as the worm wriggles away, full of energy, ready to continue its life cycle.

This is a true story, not a scene from a science fiction film. The main character is not a human being at all, though, but a cricket or grasshopper. The worm that emerges from its body cavity is a parasite called a horsehair worm, and the scene described above is enacted in many parts of the world as the horsehair worm leaves its host to reproduce in water. The worm can grow quite happily inside the cricket, having arrived there when the cricket unwittingly ate a worm egg along with its usual diet of dead leaves or other detritus, but the inside of a cricket provides a rather limited social environment, and when it is time to mate and lay eggs, the worm must seek water to find its companions. The catch is that crickets do not spend time in the water, not being aquatic animals. Nothing daunted, the worm solves the problem by a deft bit of mind control: It changes the nervous system of its host so that it seeks out an environment that it would never prefer without the parasite's influence. The exact details of how the alteration is performed are still unclear, although researchers have isolated some chemicals produced by the worm that appear to create the desired effect.

Control issues crop up in any relationship. Decisions may benefit one party more than the other, and someone usually takes the lead in determining who is dominant. Does he always decide where they vacation? Does she have the power to ban junk food from the house? The interaction between parasite and host is no exception, although the fact that each belongs to a different species, one living inside the other, introduces some complications. All behavior, whether in crickets or humans, simple or complex, comes from the brain and central nervous system. But if more than one set of genes exists inside the same body, who directs that body's nervous system? Since the worm cannot use its own muscles to get to a stream, it does something even better, by directing the muscles of its host. Pathogens change the physiology and appearance of their hosts in many ways to make their own lives easier, and it is only a small step from attaching yourself to the wall of the intestine to altering the brain running the body containing that intestine to control the host's movement.

Parasites can change the personalities of their hosts, making them more gregarious or aggressive. Like the cricket that was not sure why it found itself near the stream, other hosts, including people, can find themselves doing things out of character, driven by their diseases. And if no one is separate from disease, does this mean that our actions are ever our own? Or are they always an unwitting compromise of what we think we want to do and what would be best for our parasites?

THE LONG ARM OF THE GENE

Richard Dawkins, evolutionary biologist and author of the landmark book *The Selfish Gene,* looks at everything from the gene's perspective. For a gene to be perpetuated, it must co-opt the energy of the body it finds itself in. If the body seeks nourishment and finds a mate, the gene will survive in the offspring produced by that body, so selection will favor good food-getting and mate-finding abilities. From a parasite's gene's perspective, manipulating its host to find food is not qualitatively different from "manipulating" its own body. In both cases, the gene that replicates, whether by being in a body that successfully feeds itself or by directing the host of that body to feed, is the gene that wins.

Dawkins calls this long-reaching effect "the extended phenotype"; the phenotype is what we can observe in an organism, as opposed to its genotype, which is what genes that organism possesses. Thus there may be a gene called Xqr45 coding for knuckle hair, to give a hypothetical example, so that an individual with one form of the gene has an Xqr45A genotype and the hairy knuckle phenotype, while someone with the Xqr45a genotype has smooth, hairless knuckles. But the genes can have an extended effect in individuals whose bodies do not actually produce them, so long as the actions of those individuals increase the likelihood of the gene's replication. All genes, he says, are parasitic in their way. It's just the extent of their reach that differs.

In a more true-to-life example, a spider in the forests of Costa Rica ordinarily spends its time spinning respectable-looking orb-shaped webs to trap unwary insects. The insects feed the spider, the web aids its prey-catching ability, and all is well. But if the spider is discovered by a tiny parasitic wasp, the rules of the game change. The wasp stings the spider, paralyzing it for ten or fifteen minutes, while she glues an egg onto the spider's abdomen. The spider awakens and proceeds to live a normal arachnid life for the next couple of weeks, while the wasp egg hatches into a larva that remains attached to the spider and sucks its blood for nourishment. Then, the night before the wasp larva is ready to form a cocoon, the spider does something very different. It builds a web unlike its capture devices, instead constructing what looks like the top of a circus tent. Then the wasp larva kills and eats the spider, and forms a cocoon hanging under the web like a trapeze, protected by the spider web awning. The adult wasp eventually emerges from its shelter, ready to direct a spider of its own.

The spider's body is doing the spinning, of course, but it is all in the cause of the wasp's genes. We watch the spider, but we are seeing the intentions of the wasp. The spider is no more in control of its actions than a bulldozer operated by a construction worker.

Of course, a parasite can affect its host without manipulating the host's behavior for the parasite's benefit. If a diseased animal becomes sluggish and is more easily caught by a predator than its healthy counterpart, it does not necessarily mean that natural selection has made the pathogen change its host's sprint speed. Some changes in host behavior are simply part of the pathology of illness. Most scientists who study the behavioral effects of parasites point out that it is important to distinguish between true adaptations—changes that evolution produced in the parasite specifically to benefit its own genes—from what Dawkins calls "boring by-products." In other words, just because having a cold makes your brain feel fuzzy doesn't mean that being less able to complete the crossword puzzle is a sinister plot, anthropomorphically speaking, on the part of the cold virus to aid its spread. It could be, of course, and one could show this by proving that people with the cold spread it more easily because they, say, ask for help on 8 across and breathe on their neighbors, or put down their pencil more often, which enables others to pick up the virus from their germ-laden fingers. But these scenarios are pretty farfetched, even for the most ardent believer in the role of evolution in everyday life. It is more likely that the brain fuzz is simply a by-product of the effect of the cold on the rest of your body. Similarly, although a host that seeks food benefits its parasites, it also benefits itself, so the behavior is not generally viewed as a manipulation on the part of the parasite.

The idea that some disease symptoms help the pathogen, while others are adaptations on the part of the host to resist the effects of the disease, is familiar from the discussion of Darwinian medicine. Here, the focus is on ways the parasite might manipulate the host's own behavior in ways that facilitate the parasite's transmission, and while the boring by-product alternative always lurks nearby, plenty of other cases where the manipulation is crystal clear can be found.

THE ONE IN THE MIDDLE LOSES

Some of the best of those cases occur in parasites that have what are called complex life cycles, in which the parasite must complete different parts of its life cycle in multiple types of hosts. A kind of worm called a thorny-headed worm, for example, spends its youth inside a pillbug, that humble crustacean sometimes called a roly-poly for its ability to roll up into a tiny gray ball when threatened. But the pillbug is only a so-called intermediate host; to mature, the worm must find its way into the final host, a bird, which will harbor the adult worms and excrete the worm eggs with its own feces. The eggs are then consumed by the hapless pillbug, thus completing the cycle.

But how does a microscopic worm inside of a small trundling pillbug make its way into the bird? Birds do eat pillbugs. But while millions of birds eat millions of meals, only a tiny fraction of them would ordinarily consist of pillbugs at all, much less pillbugs with the worm inside. So the worm takes matters into its own hands, metaphorically speaking, and makes the host it inhabits especially alluring to a potential predator. Sensible pillbugs spend the day hiding under leaves or other shelter. But infected pillbugs brazenly wander on dry, light-colored surfaces, unlike their prudent kin, and end up being eaten by birds in numbers far out of proportion to the overall pillbug population.

Janice Moore, a parasite ecologist at Colorado State University, elucidated all this some years ago with a series of painstaking and elegant experiments using captive starlings. She was, however, surprised to find the entire story mentioned in the *Creation Research Society Quarterly* "as one 'evidence of design,' with an attribution to Revelation 4:11, 'for thou hast created all things and for thy pleasure they are and were created' (E. N. Smith 1984, p. 124). I can understand how parasite-induced behavioral alterations would interest a deity, and I am pleased to be in such good company, but if [the worm] was indeed especially created, that must surely seem a perversion to a starling or a [pillbug]."

Moore also discusses a peculiar attribute of another group of worms called trematodes. These produce infective stages called cercariae, tiny tailed beings that thrash in the water to locate their next host. Some species swim actively, but others are slow and awkward, forming clumps of cercariae that move as one, tails entwined. The clumps are probably easier to detect as prey, and hence more likely to be swallowed by a passing fish. These groups have the unlikely name of Rat King cercariae, or Rattenkönigcercariae, although the origin of the term is mysterious. Moore points out that rat catchers used to tie rats together by their tails, and that German folklore contains references to a king of the rats, who was really "a group of powerful rats joined at their tails and fed by other subservient rats." Why early parasitologists were moved to adopt the legend for scientific purposes is unclear.

Parasites can also influence the behavior of larger, more conspicuous prey. Moose and caribou are infected by a tapeworm called *Echinococcus granulosus,* which then must be eaten by another mammal, including wolves or humans, to complete its life cycle. Unlike many tapeworms, *Echinococcus* can reproduce asexually inside the moose, producing large cysts that can interfere with the host's activities. Parasitized animals run more slowly than uninfected ones, a difference that just might be more than a boring by-product of general ill health, because wolves are more likely to catch one of the worm-laden prey, successfully transmitting the parasite. The same subterfuge seems to work for human predators; hunters shoot a disproportionately high number of infected moose, particularly early in the hunting season. Doubtless the triumphant hunter attributed his success to his own skill at the chase, rather than the insidious efforts of a worm seeking entry to his flesh.

Other parasites make their hosts more vulnerable to predation directly, not by making them sluggish but by influencing the brain. Killifish, the most abundant fish of estuaries and salt marshes of Southern and Baja California, are infected with a worm that encysts in the brain case. Fish with the worms are over thirty times more likely than healthy fish to be eaten by birds, even though the parasitized fish appear to be perfectly healthy, and are of normal weight with normal reproductive organs. This suggests, according to Armand Kuris, a parasitologist at the University of California at Santa Barbara, "that the manipulation by the parasite is physiologically subtle and sophisticated." The worms must change some tiny but crucial part of the fish's behavior, invisible to the casual glance. Perhaps it makes the fish swim just that much slower, or pay just a little bit less attention to a looming beak. Such precise alteration of a complex behavior could provide a perfect opportunity to study the way that the brain directs behavior under normal circumstances, if the target of the worm in the brain of its host could be elucidated. Furthermore, Kuris notes, the worm itself alters the ecology of the marshes by making a food source for many bird species much more readily available than if the fish were healthy. Give a bird a fish, and it eats for a day. Infest some fish with worms, and the bird could eat for a lifetime.

DON'T BOTHER SHOOTING THE MESSENGER

Parasite manipulation can almost—almost—make you pity the poor mosquito. Mosquitoes and other biting flies transmit a cornucopia of diseases, including malaria, dengue fever, African sleeping sickness, and river blindness, all caused by microscopic parasites that slip from the flies' saliva into the bloodstream of the host during a bite. Of ten diseases designated by the World Health Organization and other international associations as being of special concern worldwide, particularly among people of less developed nations, seven are transmitted via such intermediaries, or vectors. The mosquito itself is not doing the parasite a favor; female mosquitoes and related flies require a blood meal to enable them to produce eggs, while males do not find humans or other vertebrate hosts in the least bit attractive. But the process of biting is rife with danger, given the risk of being swatted or squashed if the mosquito is detected. Mosquitoes should be selected to be cautious and stealthy in their attacks, since it is essential that they survive the bite and depart to lay their eggs. Hence they would be expected to bite only the minimum amount needed for reproduction. From the disease organism's perspective, however, the more the mosquito bites, the more hosts it can make its way into, and so a foolhardy, devil-may-care mosquito might be a more effective disease vehicle than her more discreet sister.

Jacob Koella from the Université Pierre & Marie Curie in Paris has been studying malaria and its effects on both people and mosquitoes for many years. The malaria parasite, a one-celled organism in the genus *Plasmodium,* interferes with the ability of the mosquito to draw up blood through its siphon-like mouthparts. This impairment means that the insect stabs repeatedly at its victim in an effort to obtain enough blood, increasing the chance that the *Plasmodium* itself will be transferred. Furthermore, the mosquito is more likely to become frustrated and seek a new host to top off its supply, and thus the parasite increases the chance of getting to multiple hosts in a single episode of biting. By performing detailed genetic analysis of the blood extracted from individual malaria-carrying mosquitoes from Tanzania—a feat that in my opinion puts the fictional dinosaur blood withdrawal of *Jurassic Park* to shame, along with anything the CSI techies can do—Koella and his colleagues found that 22 percent of mosquitoes infected with *Plasmodium* had bitten more than one person a night, versus just 10 percent of female mosquitoes that did not harbor the parasite.

The saliva of mosquitoes turns out to be remarkably interesting stuff. If the mosquito is just working for herself, so to speak, her saliva carries substances that facilitate her feeding, making the blood of her victims flow more readily. But infection with *Plasmodium* reduces the amount of these substances by three-quarters, making the blood much more difficult to suck up into her mouthparts. Malaria-carrying mosquitoes also lay fewer eggs than uninfected ones, which means that more resources can be left over for the use of the parasite, and some evidence suggests that they may also live longer.

Sandflies transmit a rather nasty tropical disease called leishmaniasis, in which the parasite can cause tissue in the skin or cartilage to disintegrate. The microscopic blood parasite that causes leishmaniasis manipulates the digestive system of the fly so that its gut is occluded and it keeps attempting to feed even though it is unable to ingest a full blood meal. Even as the fly struggles to get blood, the parasite swims happily through its mouthparts into the host.

Even the fleas that transmit the dreaded bubonic plague parasite, the bacteria *Yersinia pestis,* are not invulnerable to the powers of manipulation. The flea's mouthparts are also blocked by the pathogen, and the bacteria erupt into the host as the flea tries desperately to feed. Eventually, the rat or other rodent the flea resides on is killed by the bacteria, causing the flea to seek out another host and spread the bacteria further.

POSSESSED BY AN UNKNOWN RAGE

While it is unclear why Englishmen behave oddly, the reason that mad dogs go out in the noonday sun, as well as act in other uncharacteristic ways, is well known: They are being manipulated by a parasite, the rabies virus. The transmission of the virus from dogs to people was recognized over 2,000 years ago, by the Babylonians, and rabies is still a serious threat where dogs are not commonly vaccinated, killing more than 30,000 people worldwide each year. The increased tendency to salivate, causing the characteristic foaming at the mouth, occurs because the virus attacks the salivary glands and the nerves associated with swallowing, so that saliva accumulates around the mouth. Rabies is sometimes called "hydrophobia," or fear of water, because the muscle spasms associated with the virus' presence make drinking extremely painful.

In dogs and some of the other mammals that rabies can infect, the virus also has another effect; it makes the animal more aggressive and likely to bite, a behavior that transmits the virus to a new host. Interestingly, and contrary to popular belief, bats do not commonly carry rabies and when they do are unlikely to become aggressive and attack people; although handling a sick bat is unwise, worry about bats swooping down and biting innocent bystanders is unfounded. Bats also do not get into people's hair, again contrary to a common misconception—as I point out to my students, unless they have been storing insects in their hair, there is nothing on their heads that the bats could find of interest. And bats eat vast numbers of disease-carrying flies and mosquitoes, making them extraordinarily useful animals to have around.

The rabies virus differs from the parasites discussed previously because it does not have several different kinds of hosts; the same disease is transmitted directly from one animal to another, with no vector or intermediate host required. And it does not always make its host more aggressive. In dogs, rabies has a "dumb" form, in which the animal is not more likely to seek out another animal to bite, in addition to the "furious" form, in which it is. Nonetheless, the virus benefits by its host becoming more belligerent, since the virus is carried in the saliva.

Why the virus doesn't always induce aggression is an intriguing and as yet unanswered question. But when it happens, it represents a particularly extraordinary achievement. After all, aggression is a very complicated behavior—we do not seem to be able to understand its source in small boys, let alone its manifestation in war, spousal abuse, or serial homicide. So how does a virus, which not only lacks a brain, a nervous system, but even any other cellular machinery of its own, manage to do it? No one knows for sure, although several studies have isolated the nerve cells that are targeted by the virus as well as some of the chemicals whose levels the virus alters. In some cases, it seems the manipulation of host behavior arises as the parasite tries to get around the host's own immune system. Once you are mucking around in there, changes in other inflammatory responses are just a few short steps away. Robert Sapolsky, the neurobiologist from Stanford, suggests that we take advantage of the pathology of rabies to try to understand the neurobiology of aggression. It is a humbling but promising idea: Allow a microscopic pathogen to teach us more about our own minds.

The rabies virus is not creating aggression out of whole cloth, of course. It is just capitalizing on behavioral tendencies that are already there, and the seeds for this kind of manipulation can be seen in diseases other than rabies. For example, cats get both feline leukemia virus (FeLV) and feline immunodeficiency virus (FIV). FeLV has several routes of transmission, including sexual contact, biting, and grooming, but FIV is present in saliva and transmitted only via bite wounds. Male and female cats are equally likely to be infected with FeLV, but FIV is more common in males, which fight—and bite—more commonly. Does FIV make its hosts more likely to get into altercations? To my knowledge, no one has investigated this possibility, but the stage is certainly set for a rabies-like manipulation of the cat's behavior.

Wounding is the primary way that another virus, called Seoul virus, is transmitted among male rats. Sabra Klein, a scientist at the Johns Hopkins Bloomberg School of Public Health in Maryland, has studied the effects of Seoul virus, a relative of the hantavirus that has periodic outbreaks in the southwestern United States, using rats. She does not, however, use the calm, gentle laboratory rats one procures from a commercial supplier. She goes into the alleys of Baltimore and traps the mean, canny street rats of urban lore, and then she anesthetizes them (as much to protect herself as to keep the rats from feeling pain), takes blood and saliva samples, and carefully examines their fur for scars and wounds.

About half the rats of Baltimore are infected with the virus. It turns out that older males and those with more wounds were more likely to have antibodies to Seoul virus in their systems, and more likely to shed the virus in saliva, urine, and feces. Males with many wounds also had more testosterone. Female rats are generally less susceptible to the virus, as well as being less aggressive, and in the trapped females wounds were not associated with carrying or transmitting Seoul virus. Furthermore, the infection actually seems to cause the aggression in the males, rather than the behavior being a by-product of viral disease, because male laboratory rats that Klein and her colleagues experimentally infected were more likely to attack a cage intruder or exhibit a dominant posture. This behavioral change occurred only while the virus was more likely to be transmitted, not during the early phase of infection when the rat is actually the sickest.

Interestingly, viruses seem more likely than other disease-causing organisms to make their hosts more aggressive; tick-borne encephalitis, for example, is also said to make its rodent hosts more likely to fight, and the herpes simplex virus is associated with higher levels of aggression, at least in mice. There is

even a virus said to make bees more hostile. Whether this is because viruses, being incorporated into their host's cells, have better access to the machinery of the brain and nervous system
remains to be seen.

MANIPULATING THE MANIPULATOR

The parasite does not always succeed in its manipulation, although in at least one case it is subverted not by the host overcoming its wiles but by yet another parasite. A New Zealand
cockle ordinarily lives a modest molluscan life in intertidal sand
and mudflats buried a few inches below the surface. The parasite,
a type of trematode worm, leaves its first intermediate host, a
snail, and enters the cockle through the siphon the cockle uses as
a breathing tube. Once inside the cockle, the worm travels to the
foot, the meaty muscle that protrudes outside the shell. As the
worms accumulate in the cockle's foot, it becomes less and less
able to burrow into the sand, and instead lies exposed on the surface of the mud, where it is more likely to be discovered by a
predatory bird, the final host of the trematode.

But the presence of the cockles on the surface does not go
unnoticed by other, unintended recipients. Robert Poulin from
the University of Otago in New Zealand, one of the most influ-
ential scientists studying parasite manipulation, discovered that
the cockle feet were nibbled upon in such a way that the cysts of
the worms were consumed, leaving the bivalve itself relatively
unharmed. These foot croppers, as he and his coworkers called
them, turned out to be a fish that took advantage of the unusual
availability of the cockle foot to get an easy meal. Because the fish
is not the final host of the worm, this behavior thwarts the parasite's manipulation. Poulin later determined that the surface rest-
ing behavior of the cockle does not necessarily enhance the

parasite's transmission by as much as would be necessary to be a true manipulation rather than one of those boring by-products of infection. Furthermore, birds often choose prey based on size, and they may avoid the largest cockles, which are likely to carry the heaviest parasite loads. Nevertheless, from the worm's perspective, ending up inside the stomach of a fish rather than the rightful bird host is a dead end, and the story suggests that even the best-laid plans of worms can go astray.

GIANTS, EUNUCHS, AND WHY LIVING LONGER IS NOT ALWAYS A GOOD THING

If the cockle's tale is a lesson in parasites not always getting their way, the examples that follow are the reverse, where what looks like a host escape is just more insidious manipulation by the parasite. Many parasitology textbooks contain a photograph of two mice, one infected with a tapeworm and one free of parasites. One mouse is skinny, the other large and rotund, like a toy, looking like a fur ball with a tail attached. Surprisingly, the plump mouse is the one with the worm, while its runty cousin is healthy. The worm apparently secretes a chemical that mimics growth hormones, increasing the size of its host. Being bigger is generally beneficial for animals, but in this case and others like it, the parasite-induced gigantism, as it is termed, is actually better for the parasite. The tapeworm in the mouse needs to get into a final host, and what better prey than one that is too fat to easily escape its predators? In effect, the worm is using the mouse itself to construct a better mousetrap.

In other cases, gigantism is more a matter of making the host's body into a better incubator for parasites, and a bigger host may well provide exactly that. People who eat and eat without gaining weight can joke that they have a tapeworm, but really, who has the last laugh? Some researchers maintain that the gigantism is actually a host defense mechanism, enabling greater resistance to the effects of the parasite. However, in at least one other case besides the mouse and tapeworm (a snail infected with a trematode), the parasite again is responsible for producing substances that direct the increased growth rate, suggesting that it is the parasite's interests that are favored.

From a parasite's perspective, the host shouldn't do anything risky or difficult, unless such an activity promotes transmission, as in the situations cited previously in which intermediate hosts need to be eaten by another animal to complete the parasite's life cycle. Otherwise, however, the parasite is best off if its host takes it easy, not engaging in any demanding activities. Given that the most demanding thing most organisms do is reproduce, it stands to reason that a parasite that can keep its host from reproducing will be able to use the resources and energy its host would otherwise have channeled into making offspring. A celibate host is a better host, and a host that not only doesn't reproduce but can't is better still. And so many parasites, particularly those in crustaceans like crabs and shrimp but also some in other groups, evolved a means to castrate their hosts.

Castration is something of a misnomer, at least given the colloquial use of the word to refer to the removal of male reproductive organs, since both male and female hosts can be affected. Several parasites of crabs, for instance, take advantage of the protected location on the abdomen where a female's eggs would normally be brooded, instead sending out tendrils of tissue and replacing the crab's ovaries with their own organs. The parasite then enjoys the shelter and maternal behavior of the female crab, who solicitously guards her "offspring." Since she never actually has to marshal the resources necessary to manufacture real eggs, however, the crab often lives longer or grows larger than her fertile counterparts, which is all to the good of the parasite. A barnacle that parasitizes both male and female crabs feminizes the males first, and then mimics the reproductive organs of a female crab in both sexes of host; the host then goes into deeper water and fans the fake eggs as if they were truly her—or his—own. Some parasites induce their hosts to undergo "false oviposition," in which the host goes through all the motions of egg-laying only to produce eggs or larvae of the parasite instead.

Parasitic castration often makes the host keep a juvenile appearance long after it should have matured, so that it is like a giant baby, growing large but never developing the secondary sexual characteristics used in mate attraction or other aspects of reproduction. Parasitized male cockroaches seem normal, but are indifferent to the odors produced by sexually receptive females, odors that ordinarily would drive a male cockroach wild. In the case of a hermaphroditic host, where both male and female sexual organs are present in the same individual, castrating parasites may affect one sex but not the other; a liver fluke that uses snails as an intermediate host, *Schistosoma mansoni,* halts egg production in its host but does not seem to alter the ability of the same individual to produce sperm.

Curtailing reproduction, or at least changing its timing, might also be beneficial to the host. It can serve as a kind of damage control, so that a host already compromised by having to deal with the inroads of infection does not also invest in costly reproduction. Alternatively, several parasitologists have suggested that it behooves a host that can sense it has been infected to immediately throw all the switches and reproduce as quickly as possible, before its resources are ravaged by disease. Even if reproduction is diminished overall, the host will have made the best of a bad job.

Being rendered completely sterile is a fate that, if not worse than death, is just as bad, evolutionarily speaking. What is certainly worse than the death of an individual, however, is the death of the same genes in other individuals. In other words, if an infected animal not only stops reproducing but infects all its relatives, the genes of that animal, which of course are also represented in its kin, suffer accordingly. So perhaps, if you are already doomed, it would be better to simply kill yourself and spare your kin the same fate. That way, your genes might still be perpetuated indirectly.

Biologists have applied this indirect genetic benefit idea to unselfish behavior in general. It is called kin selection; the details are beyond the scope of this chapter, but they are related to the gene's-eye view championed by Dawkins. In the case of parasite-induced suicide, several scientists have proposed that if infected animals are surrounded by kin who share their genes and to whom they are likely to transmit the parasite if they hang around, or who will bear a cost if they linger, natural selection should favor killing oneself early on in the disease process.

That is precisely what a group of Canadian scientists suggested was happening in an aphid that was parasitized by a wasp. Aphids, you may recall, can reproduce asexually, so that they are likely to be surrounded by dozens of identical twin sisters; even when this is not the case, the likelihood that nearby aphids share many genes is quite high. The wasp lays its eggs in the aphid, and eventually causes its host to shrivel up and mummify. If the eggs go into a young aphid, it will die before reproducing, but an older aphid has a chance at having some offspring before the effects of its parasite take over. Accordingly, the scientists found that the juveniles were far more likely to drop off the plant when approached by a predator, virtually guaranteeing their death, while the older aphids were more prudent in their responses. The claim that aphids were committing suicide was met with considerable skepticism by other biologists, with some suggesting that the response is simply part of the pathology of the wasp affecting the ability of aphids to move around. Nevertheless, it is an intriguing idea, and one that is at least theoretically reasonable.

Similar adaptive suicide has been suggested to occur in bumblebees and other social insects, which live in complex groups of close relatives. Here the idea is that sacrificing oneself is beneficial for the colony if the parasitized bee cannot do its job well and is merely living off the efforts of its more industrious colony mates. The melodramatic statement that "You would all be better off if I were dead!" might be met in such cases with measured agreement.

SEX, PAIN, AND PARASITES

Everyone is familiar with the idea that extreme stress can make people ignore pain: the athlete who turns out to have run half a marathon on a fractured ankle, the soldier who realizes he has a severe wound only after the battle. Parasites too can alter the pain thresholds of their hosts, for reasons that still remain tantalizingly unclear but that illustrate once again the blurred boundaries between what favors the host and what favors the disease.

Martin Kavaliers of the University of Western Ontario in Canada has spent many years studying the ways that mice detect parasites in other individuals and how they respond when they do so. Because one cannot ask a mouse to report how its perception of pain changes under different circumstances, Kavaliers uses a simple but ingenious technique my husband dubbed "mouse on a hot tin roof." The rodent is placed on a metal plate that can be heated just until it becomes uncomfortable, like the sand at the beach on a summer day. Like the swimmers going to the shore, the mouse lifts its foot or tail off the heated surface after it gets too hot, and the more sensitive it is to pain, the more quickly it does this. The greater the number of seconds until liftoff, the higher the pain threshold is on the part of the animal.

When Kavaliers and his coworkers infected mice with a one-celled parasite, both male and female mice were less sensitive to pain. Interestingly, healthy female mice allowed to sniff bedding from a parasitized male were less sensitive to pain than those sniffing the odors of healthy males, and they would rather mate with the latter when given an opportunity. The preference for parasite-free mates is not surprising, as I discussed in an earlier chapter, but the relationship to the pain response is more enigmatic. Furthermore, parasitized male mice were indifferent to the odor of a sexually receptive female when the parasite was in its early stages and not infective, but as the infection developed, the mice began to show a heightened interest in the female odors. Male mice also avoid the odors of parasitized females, and exhibit less pain sensitivity when exposed to them.

Kavaliers suggests that the change in pain response may help the animal avoid distraction during crucial times, and this may be the case, or it could simply be a by-product of a stress response. The change in male preference for receptive females when the infection reaches its height suggests parasite manipulation, and the change in pain sensitivity could be irrelevant, a side effect of the place in the nervous system that the pathogen affects. Or perhaps the reduction in pain is a host defense that makes coping with disease easier. Is pain part of the control issue in our relationship with disease?

Even more significantly, parasitized mice also spent longer near the odor of a predator, which leads us to the best example yet of parasites influencing not merely a few aspects of behavior, but the entire personality, and in humans as well as mice.

SORRY, OFFICER, I HAVE PARASITES

Like many of the parasites we have met, toxoplasmosis is transmitted from its intermediate host to its final host by predation: A rodent acquires the infection from soil, and is eaten by a cat or other predator, where the one-celled parasite completes its life cycle, its own reproductive stages excreted along with the cat's feces.

This is all well and good, but rats and other rodents have a long evolutionary history of being eaten by cats, and selection has produced in rats the sensible response of avoiding anything that seems to have been in contact with the predators, including the odor of places that cats frequent. Even rats kept in the laboratory for several hundred generations possess an atavistic wariness of things feline, regardless of whether they themselves have ever been in any environment other than a small plastic cage supplied with food pellets and a water bottle. So the toxoplasmosis organism does what other parasites do, and alters the behavior of the host so that it is likely to be more vulnerable to being eaten.

Using laboratory-reared rats whose great-great-grandparents had been caught on farms in England, Oxford University researchers Manuel Berdoy and Joanne Webster set up a maze with a wooden nest box in each corner. The straw in the nest boxes was scented either with the rats' own smell, rabbit urine, cat urine, or water, to serve as a neutral stimulus. Some of the rats were infected with toxoplasmosis, and others were kept as controls. Each rat was released into the maze at dusk and its behavior recorded with a video camera.

The uninfected rats showed what Berdoy and Webster called "a healthy aversion of cat-scented areas," while the parasitized rodents frolicked with abandon in them, apparently oblivious to the signs of danger. Neither group of rats cared one way or the other about the neutral or rabbit odors, and both were happy to spend time in the nest boxes smelling comfortably of rat. Other behaviors, such as aggression, remained normal; only the response to cats was altered. Watching the video of parasitized rats wandering haplessly into the boxes with eau de cat was like seeing the heroine in a horror movie open the door to the deserted barn while the maniac with the ax lurks behind it.

This reckless behavior is clearly advantageous to the parasite, which is much more likely to be transmitted in foolhardy prey, just as the thorny-headed worm altered the preference of a pillbug for well-lit open spaces. The careless disregard for areas frequented by cats may also increase the likelihood of the rodents picking up the parasite from contaminated soil to begin with, as well as the chance of becoming prey.

People, as Webster laconically observes, "are rarely preyed upon by cats," but traces of infection with the toxoplasmosis parasite are found in anywhere from 22 percent to 84 percent of humans, depending on the country examined. Undercooked meat can be a major source of the parasite in places where regulation of agricultural products is less than stringent. The disease itself has few overt manifestations in people, and unless a pregnant woman contracts it, as discussed previously, it is usually not a matter for medical concern.

But the signs of the parasite in humans may be at least as subtle as the ever-so-slightly slowed reflexes of the killifish. Jaroslav Flegr from Charles University in Prague has been examining the differences in personality between people showing signs of previous infection with toxoplasmosis and those who are parasite-free. He started by administering a standard personality test to 338 students and university staff, both men and women. The women showed no effect of the parasite, but infected men were more reserved and less trusting of others, with a greater propensity for disregarding rules. The difference was small, but distinct, and a follow-up study using a larger sample showed similar results in men and also a tendency for women harboring signs of toxoplasmosis to be more outgoing, trusting, and self-assured.

Flegr then moved to a more heterogeneous sample, if only of one sex, and gave the same personality test to 857 military conscripts. Here the infected subjects scored lower in their tendency to seek novelty, which fits in with the less trusting nature of the men in the earlier survey, and somewhat surprisingly also had a slightly lower IQ. He and his coworkers also examined the reaction times of both men and women, with a commonly used test in which the subject has to press a key when a dark square appears on a computer screen. Infected people had slightly longer reaction times, a response that seems consistent with the effect of toxoplasmosis on rodents; slower reaction times would presumably make a rat easier prey.

In people, reaction time may be important in another context. In an ingenious comparison of 146 Prague residents involved in traffic accidents with 446 people living in the same places where the accidents were reported, Flegr and his colleagues found that the signs of toxoplasmosis were much more common in the accident group. This was true regardless of whether the person was a pedestrian who had been hit by a car or was driving, and those who could not have influenced their accident in any way were excluded from analysis; in other words, a victim was not tested if, say, his or her car was hit from behind while stopped at a traffic light. Those with more recent signs of infection had a higher accident risk than people whose infections had been present for a long time. While Flegr does not suggest parasite infection as an excuse to get out of a ticket, one wonders if it is only a matter of time until someone does, particularly in more litigious countries than the Czech Republic.

So does toxoplasmosis alter people's personalities, or are people with certain personality types more likely to contract the disease? Although some researchers have suggested that the association between mental difficulties and toxoplasmosis is due to poor hygiene among those with psychological disorders, this seems unlikely given the variety of routes for acquiring the disease; it is hard to argue that disturbed individuals are more likely to buy contaminated meat, or to garden in areas frequented by cats. The growing understanding of the way the parasite targets the brain also suggests that the alterations occur after infection, since these provide a mechanism for the parasite's manipulation. Even though humans are not the intended host of the parasite, it still influences the same brain regions that make the rats more vulnerable prey.

It is, however, possible that certain people with genetic predispositions toward particular personality types also have a greater susceptibility to toxoplasmosis. Flegr's group also found that men who tested positive for toxoplasmosis tended to be a bit taller, and the relative lengths of their second and fourth fingers were more similar. These traits are associated with higher levels of prenatal testosterone, and we have already seen that testosterone influences disease susceptibility in males from a variety of species. Whether the personality differences between uninfected and infected individuals that Flegr found hold up in other cultures is as yet unknown. Nonetheless, it would be extremely interesting to see whether the sex distinctions, for instance, hold for cultures in which sex roles differ from those in Prague.

LISTENING TO PARASITES

If women with toxoplasmosis are more "warmhearted," as Flegr puts it, or men are more prone to break rules, does this mean that kindness is always a personality characteristic, or is it sometimes a symptom of disease? If curing people of an illness, whether toxoplasmosis or anything else, would change their temperament, but that illness has been a chronic part of them for many years, is their personality even their own? No one denies that traits like recklessness or kindness are affected by a person's environment, but what if one woman is also influenced by her parasites? Is she any less kind than the woman whose nature is warm without the benefit of a pathogen? Who are we, really, but the sum of our own cells and those of our parasites, intertwined in a relationship that will never end? The effect of toxoplasmosis in rodent brains is similar to that of some antianxiety drugs. Just as Sapolsky suggested that we try to understand the biology of aggression by studying the neurobiology of rabies, perhaps we could gain insight into anxiety and worry by examining toxoplasmosis. How are the parasites' effects different from Prozac's? And why is administering the latter acceptable, while we strive to eliminate the former?

Our close relationship with disease makes us ask some uncomfortable questions about who we are. Kevin Lafferty from the University of California at Santa Barbara even suggests that the stereotypical differences among cultures—the outgoing Latin Americans, the subdued Scandinavians, the orderly Germans—could be due to longstanding differences in infection. People from different places get different diseases, and diseases can influence personality. Again, few would argue that the pathogen alone dictates our essence, but then few would argue that we are solely the product of any one factor, whether it be our genes, our hormones, our birth order, or our early experiences with toilet training. Personality is the sum of all of these things, and it seems artificial to disregard parasites' influence.

One could draw an analogy here with deafness, which many people would unquestionably classify as pathology but which is also the source of an entire culture, with its own language and social system. Some deaf activists resist the idea of "curing" deafness with cochlear implants, because they say that this implies deaf people are defective, rather than simply being variations within the enormous diversity of human beings. If we are shaped by our diseases, then removing those diseases might not only disrupt our immune systems, but our personalities might change. It's unclear whether it is more natural to live with those changes, even if another organism's DNA has affected them.

There is no question about whether curing severe illnesses like plague or flu or acute psychosis is warranted. But we have a cadre of more subtle hangers-on, from eyelash mites to pinworms to bacteria floating in our blood. If diseases have always shaped us, then the problem (if problem it is) of how changing our bodies will change our minds goes much deeper than we had thought. I am inclined to give at least some of our internal inhabitants a bit of mercy. But then, perhaps I am not fully in control of my inclinations.

ACKNOWLEDGMENTS

My interest in parasites and disease goes back at least to my time as a graduate student at the University of Michigan, where Bill Hamilton was my advisor. Bill saw the effects of parasites everywhere, and encouraged me to do the same. He died in 2000, but I hope he would have appreciated even the most outlandish speculations in these pages; in fact, he might have enjoyed those most of all. Ed Platzer has always been a mentor and wellspring of fabulous parasite facts, and many other scientists, including Beverly Strassmann, Randy Nesse, Mike Worobey, Dan Brooks, Charlie Nunn, Curt Lively, Peter Timms, Peter Hudson, Rob Knell, and Joel Weinstock offered unpublished manuscripts, advice, and references. I am, as always, amazed and touched by the generosity of my colleagues. Randy, Mike, Charlie, and Curt also each vetted one or more chapters and steered me away from errors, though any remaining mistakes are of course my own. John Rotenberry continues to provide numerous forms of support. David Edwards read every chapter and made many helpful suggestions, and Deborah Blum served as a sounding board and source of information about writing and publishing. Several chapters were written while I was on sabbatical at the University of Western Australia, and I am grateful for the hospitality and support I received there, particularly from my longtime collaborator and friend Leigh Simmons. Many thanks also to my agent, Wendy Strothman, who patiently led a novice through the commercial publishing process, and to my extraordinary editor, Andrea Schulz, who has an uncanny ability to offer criticism without injuring the ego.

REFERENCES

CHAPTER 1: WHY DOCTORS NEED DARWIN

Berlim, M. T., and Abeche, A. M. 2001. Evolutionary approach to medicine. *Southern Medical Journal* 94: 26–32.

Blanford, S., Thomas, M. B., and Langewald, J. 1998. Behavioural fever in the Senegalese grasshopper, *Oedaleus senegalensis,* and its implications for biological control using pathogens. *Ecological Entomology* 23: 9–14.

Blatteis, C. M. 2003. Fever: Pathological or physiological, injurious or beneficial? *Journal of Thermal Biology* 28: 1–13.

Eaton, S. B., Strassmann, B. I., Nesse, R. M., Neel, J. V., Ewald, P. W., Williams, G. C., Weder, A. B., Eaton, S. B. III, Lindeberg, S., Konner, M. J., Mysterud, I., and Cordain, L. 2002. Evolutionary health promotion. *Preventive Medicine* 34: 109–118.

Eaton, S. B., Cordain, L., and Lindeberg, S. 2002. Evolutionary health promotion: A consideration of common counterarguments. *Preventive Medicine* 34: 119–123.

Elliot, S. L., Blanford, S., and Thomas, M. B. 2002. Host-pathogen interactions in a varying environment: Temperature, behavioural fever and fitness. *Proceedings of the Royal Society of London* B 269: 1599–1607.

Elliot, S. L., Blanford, S., Horton, C. M., and Thomas, M. B. 2003. Fever and phenotype: Transgenerational effect of disease on desert locust phase state. *Ecology Letters* 6: 830–836.

Havinga, W. 2003. Time to counter "fever phobia"! *The British Journal of General Practice,* March, 253.

Hofbauer, K. G. and Huppertz, C. 2002. Pharmacotherapy and evolution. *Trends in Ecology and Evolution* 17: 328–334.

Kluger, M. J., Kozak, W., Conn, C. A., Leon, L. R., and Soszynski, D. 1996. The adaptive value of fever. *Infectious Disease Clinics of North America* 10: 1–20.

Kluger, M. J., Kozak, W., Conn, C. A., Leon, L. R., and Soszynski, D. 1998. Role of fever in disease. *Annals of the New York Academy of Sciences* 856: 224–233.

Kluger, M. J. 2002. Fever in acute disease—Beneficial or harmful? *Wiener Klinische Wochenschrift* 114: 73–75.

LeGrand, E. K. 2000. Implications of early apoptosis of infected cells as an important host defense. *Medical Hypotheses* 54: 591–596.

LeGrand, E. K. 2000. Why infection-induced anorexia? The case for enhanced apoptosis of infected cells. *Medical Hypotheses* 54: 597–602.

LeGrand, E. K., and Brown, C. C. 2002. Darwinian medicine: Applications of evolutionary biology for veterinarians. *Canadian Veterinary Journal* 43: 556–559.

McCrone, J. 2003. Darwinian medicine. *The Lancet Neurology* 2: 516.

Moore, J., and Freehling, M. 2002. Cockroach hosts in thermal gradients suppress parasite development. *Oecologia* 133: 261–266.

Nesse, R. M. 2001. How is Darwinian medicine useful? *Western Journal of Medicine* 174: 358–360.

Nesse, R. M. 2001. The smoke detector principle. *Annals of the New York Academy of Sciences* 935: 75–85.

Nesse, R. M. 2001. Medicine's missing basic science. *The New Physician,* December, 8–10.

Nesse, R. M., and Williams, G. C. 1991. The dawn of Darwinian medicine. *The Quarterly Review of Biology* 66: 1–22.

Nesse, R. M., and Williams, G. C. 1994. *Why We get Sick: The New Science of Darwinian Medicine.* Times Books, New York.

Nesse, R. M., and Williams, G. C. 1997. Evolutionary biology in the medical curriculum—what every physician should know. *BioScience* 47: 664–666.

Nesse, R. M., and Williams, G. C. 1998. Evolution and the origins of disease. *Scientific American* 279: 86–96.

Paul, R. E. L., Arley, F., and Robert, V. 2003. The evolutionary ecology of *Plasmodium. Ecology Letters* 6: 866–880.

Rotherman, S. 2004. Diabetes: Heading for trouble. *The Weekend Australian,* July 17–18.

Russell, F. M., Shann, F., Curtis, N., and Mulholland, K. 2003. Evidence on the use of paracetamol in febrile children. *Bulletin of the World Health Organization* 81: 367–374.

Sherman, E., and Stephens, A. 1998. Fever and metabolic rate in the toad *Bufo marinus. Journal of Thermal Biology* 16: 49–52.

Stearns, S. C., ed. 2001. *Evolution in Health and Disease.* Oxford University Press, Oxford.

Stearns, S. C., and Ebert, D. 2001. Evolution in health and disease: Work in progress. *The Quarterly Review of Biology* 76: 417–432.

Trevethan, W. R., Smith, E. O., and McKenna, J. J., eds. 1999. *Evolutionary Medicine.* Oxford University Press, Oxford.

Weiner, H. 1998. Notes on an evolutionary medicine. *Psychosomatic Medicine* 60: 510–520.

World Health Organization, Programme for the Control of Acute Respiratory Infections. 1993. The management of fever in young children with acute respiratory infections in developing countries. www.who.int/child-adolescent-health/New_Publications.

CHAPTER 2: **FRIENDLY WORMS AND THE PRICE OF VICTORY**

Anderson, A. D., Nelson, J. M., Rossieter, S., and Angulo, F. J. 2003. Public health consequences of the use of antimicrobial agents in food animals in the United States. *Microbial Drug Resistance* 9: 373–379.

Andersson, D. I., and Levin, B. R. 1999. The biological cost of antibiotic resistance. *Current Opinion in Microbiology* 2: 489–493.

BBC News, December 3, 2003. Eat worms—feel better.

Braun-Fahrländer, C., Riedler, J., Herz, U., Eder, W., Waser, M., Grize, L., Maisch, S., Carr, D., Gerlach, F., Bufe, A., Lauener, R. P., Schierl, R., Renz, H., Nowak, D., and von Mutius, E. 2002. Environmental exposure to endotoxin and its relation to asthma in school-age children. *The New England Journal of Medicine* 347: 869–877.

Brody, J. E. 2005. When trouble hits those holes in your head. *New York Times,* March 15.

Centers for Disease Control and Prevention Web site on antimicrobial resistance: http://www.cdc.gov/ncidod/aip/research/ar.html.

Cortese, A. 2005. An arsenal of sanitizers for a nation of germophobes. *New York Times,* February 27.

Eggesbø, M., Botten, G., Stigum, H., Nafstad, P., and Magnus, P. 2003. Is delivery by cesarean section a risk factor for food allergy? *Journal of Allergy and Clinical Immunology* 112: 420–426.

Elliott, D. E., Summers, R. W., and Weinstock, J. V. 2005. Helminths and the modulation of mucosal inflammation. *Current Opinion in Gastroenterology* 21: 51–58.

Furrie, E., Macfarlane, S., Kennedy, A., Cummings, J. H., Walsh, S. V., O'Neil, D. A., and Macfarlane, G. T. 2005. Synbiotic therapy (*Bifidobacterium longum/* Synergy 1) initiates resolution of inflammation in patients with active ulcerative colitis: A randomised controlled pilot trial. *Gut* 54: 242–249.

Grenet, K., Guillemot, D., Jarlier, V., Moreau, B., Dubourdieu, S., Ruimy, R., Armand-Lefevre, L., Bau, P., and Andremont, A. 2004. Antibacterial resistance, Wayampis Amerindians, French Guyana. *Emerging Infectious Diseases* 10: 1150–1153.

Hentschel, U., Dobrindt, U., and Steinert, M. 2003. Commensal bacteria make a difference. *Trends in Microbiology* 11: 148–150.

Hoesl, C. E., and Altwein, J. E. 2004. The probiotic approach: An alternative treatment option in urology. *European Urology* 47: 288–296.

Horner, A. A., and Raz, E. 2003. Do microbes influence the pathogenesis of allergic diseases? Building the case for Toll-like receptor ligands. *Current Opinion in Immunology* 15: 614–619.

Hunter, M. M., and McKay, D. M. 2004. Review article: Helminths as therapeutic agents for inflammatory bowel disease. *Alimentary Pharmacology and Therapeutics* 19: 167–77.

Inflammatory Bowel Disease Patient Community. 2000–2001. Interview with Dr. Joel Weinstock. http://Ibd.patientcommunity.com/features/weinstock_prn.html.

Isolauri, E. 2001. Probiotics in human disease. *American Journal of Clinical Nutrition* 73 (Suppl.): 1142S–1146S.

Kalliomäki, M., Salminen, S., Arvilommi, H., Kero, P., Koskinen, P., and Isolauri, E. 2001. Probiotics in primary prevention of atopic disease: A randomized placebo-controlled trial. *The Lancet* 357: 1076–1079.

Kalliomäki, M., Salminen, S., Poussa, T., Arvilommi, H., and Isolauri, E. 2003. Probiotics and prevention of atopic disease: 4-year follow-up of a randomised placebo-controlled trial. *The Lancet* 361: 1869–1871.

Kemp, A., and Björkstén, B. 2003. Immune deviation and the hygiene hypothesis: A review of the epidemiological evidence. *Pediatric Allergy and Immunology* 14: 74–80.

Kotulak, R., Gorner, P., and Becker, R. 2004. Doctors have pieces, but disease still puzzles. *Chicago Tribune,* September 26.

Larson, E. 2002. The "Hygiene Hypothesis": How clean should we be? *American Journal of Nursing* 102: 81–89.

Levin, B. R. 2004. Noninherited resistance to antibiotics. *Science* 305: 1578–1579.

Levin, B. R., and Bonten, M. J. M. 2004. Cycling antibiotics may not be good for your health. *Proceedings of the National Academy of Sciences,* U.S.A. 101: 13101–13102.

Levy, S. B., and Marshall, B. 2004. Antibacterial resistance worldwide: Causes, challenges and responses. *Nature Medicine Supplement* 10: S122–S129.

Liu, A. H. 2002. In my opinion—interview with the expert. *Pediatric Asthma, Allergy & Immunology* 15: 227–231.

Liu, A. H., and Murphy, J. R. 2003. Hygiene hypothesis: Fact or fiction? *Journal of Allergy and Clinical Immunology* 111: 471–478.

Macfarlane, G. T., and Cummings, J. H. 1999. Probiotics and prebiotics: Can regulating the activities of intestinal bacteria benefit health? *BioMedical Journal* 318: 999–1003.

Macfarlane, G. T., and Cummings, J. H. 2002. Probiotics, infection and immunity. *Current Opinion in Infectious Diseases* 15: 501–506.

McCarthy, S. 2000. Bring on the germs. Salon.com, May.

Mohamadzadeh, M., Olson, S., Kalina, W. V., Ruthel, G., Demmin, G. L., Warfield, K. L., Bavari, S., and Klaenhammer, T. R. Lactobacilli activate human dendritic cells that skew T cells toward T helper 1 polarization. 2005. *Proceedings of the National Academy of Sciences, U.S.A.* 102: 2880–2885.

Moreels, T., and Pelckmans, P. A. 2005. Gastrointestinal parasites: Potential therapy for refractory inflammatory bowel diseases. *Inflammatory Bowel Diseases* 11: 178–184.

Mottet, C., and Michetti, P. 2005. Probiotics: Wanted dead or alive. *Digestive and Liver Disease* 37: 3–6.

National Institute of Allergy and Infectious Diseases. 2004. Fact Sheet: The problem of antibiotic resistance. www.niaid.nih.gov/factsheets/antimicro.htm.

Perry, P. 2004. Worm therapy: A new treatment for IBD? *Saturday Evening Post,* July–August.

Ramsey, C. D., and Celedon, J. C. 2004. The hygiene hypothesis and asthma. *Current Opinion in Pulmonary Medicine* 11: 14–20.

Rinne, M. M., Gueimonde, M., Kalliomäki, M., Hoppu, U., Salminen, S., and Isolauri, E. 2005. Similar bifidogenic effects of prebiotic-supplemented partially hydrolyzed infant formula and breastfeeding on infant gut microbiota. *FEMS Immunology and Medical Microbiology* 43: 59–65.

Roach, M. 2004. Germs, germs everywhere. Are you worried? Get over it. *New York Times,* November 9.

van den Biggelaar, A. H. J., van Ree, R., Rodrigues, L. C., Lell, B., Deelder, A. M., Kremsner, P. G., and Yazdanbakhsh, M. 2000. Decreased atopy in children infected with *Schistosoma haematobium*: A role for parasite-induced interleukin-10. *The Lancet* 356: 1723–1727.

von Mutius, E., Pearce, N., Beasley, R., Cheng, S., von Ehrenstein, O., Björkstén, B., and Weiland, S. 2000. International patterns of tuberculosis and the prevalence of symptoms of asthma, rhinitis, and eczema. *Thorax* 44: 449–453.

Weinstock, J., and Summers, W. R. 2000–2001. Will helminths become the future treatment for inflammatory bowel disease? *Currents:* 2.

Wickelgren, I. 2004. Can worms tame the immune system? *Science* 305: 170–171.

Wills-Karp, M., Santeliz, J., and Karp, C. L. 2001. The germless theory of allergic disease: Revisiting the hygiene hypothesis. *Nature Reviews Immunology* 1: 69–75.

World Health Organization. 2000. Overcoming antimicrobial resistance, Chapter 3: Factors contributing to resistance. http://www.who.int/infectious-disease-report/2000/.

CHAPTER 3: **NOT SUCH A BAD CASE**

Best, S. M., and Kerr, P. J. 2000. Coevolution of host and virus: The pathogenesis of virulent and attenuated strains of myxoma virus in resistant and susceptible European rabbits. *Virology* 267: 36–48.

Brown, M. J. F., Schmid-Hempel, R., and Schmid-Hempel, P. 2003. Strong context-dependent virulence in a host-parasite system: Reconciling genetic evidence with theory. *Journal of Animal Ecology* 72: 994–1002.

CSIRO Australia. 1998. Rabbits: Chainsaws of the Outback. www.csiro.au/promos/billiondind/contents/rabbits.htm.

Ewald, P. W. 1995. The evolution of virulence: A unifying link between parasitology and ecology. *Journal of Parasitology* 81: 659–669.

Fenner, F. 2000. Adventures with poxviruses of vertebrates. *FEMS Microbiology Reviews* 24: 123–133.

Galvani, A. P. 2003. Epidemiology meets evolutionary ecology. *Trends in Ecology and Evolution* 18: 132–139.

Levin, B. R., Lipsitch, M., and Bonhoeffer, S. 1999. Population biology, evolution, and infectious disease: Convergence and synthesis. *Science* 283: 806–809.

Levin, B. R., and Antia, R. 2001. Why we don't get sick: The within-host population dynamics of bacterial infections. *Science* 292: 1112–1115.

Mackinnon, M. J., and Read, A. F. 2004. Immunity promotes virulence evolution in a malaria model. *PLoS Biology* 2: e230.

Schjørring, S., and Koella, J. C. 2003. Sub-lethal effects of pathogens can lead to the evolution of lower virulence in multiple infections. *Proceedings of the Royal Society of London B* 270: 189–193.

Smith, G. 1998. Looking backwards at pests. Newsletter of the Sporting Shooters Association of Australia. December.

van Baalen, M., and Sabelis, M. W. 1995. The scope for virulence management: A comment on Ewald's view on the evolution of virulence. *Trends in Microbiology* 3: 414–416 (response from Ewald immediately following).

Zimmer, C. 2003. Taming pathogens: An elegant idea, but does it work? *Science* 300: 1362–1364.

CHAPTER 4: **THE RACE WITH SEX THAT'S NEVER WON**

Bell, G. 1982. *The Masterpiece of Nature.* University of California Press, Berkeley.

Corley, L. S., Blankenship, J. R., and Moore, A. J. 2001. Genetic variation and asexual reproduction in the facultatively parthenogenetic cockroach *Nauphoeta cinerea:* Implications for the evolution of sex. *Journal of Evolutionary Biology* 14: 68–74.

Hakoyama, H., and Iwasa, Y. 2004. Coexistence of a sexual and an unisexual form stabilized by parasites. *Journal of Theoretical Biology* 226: 185–194.

Hanley, K. A., Fidher, R. N., and Case, T. J. 1995. Lower mite infestations in an asexual gecko compared with its sexual ancestors. *Evolution* 49: 418–426.

Howard, R. S., and Lively, C. M. 2003. Opposites attract? Mate choice for parasite evasion and the evolutionary stability of sex. *Journal of Evolutionary Biology* 16: 681–689.

Hurst, L. D., and Peck, J. R. 1996. Recent advances in understanding of the evolution and maintenance of sex. *Trends in Ecology and Evolution* 11: 46–52.

Jokela, J., and Lively, C. M. 1995. Parasites, sex and early reproduction in a mixed population of freshwater snails. *Evolution* 49: 1268–1271.

Kearney, M. R. 2003. Why is sex so unpopular in the Australian desert? *Trends in Ecology and Evolution* 18: 605–606.

Kumpulainen, T., Grapputo, A., and Mappes, J. 2004. Parasites and sexual reproduction in psychid moths. *Evolution* 58: 1511–1520.

Lenski, R. E. 1999. A distinction between the origin and maintenance of sex. *Journal of Evolutionary Biology* 12: 1034–1035.

Lively, C. M. 1996. Host-parasite coevolution and sex. *BioScience* 46: 107–114.

Maynard Smith, J. 1989. *Did Darwin Get It Right?* Chapman and Hall, New York.

Maynard Smith, J. 1978. *The Evolution of Sex.* Cambridge University Press, Cambridge.

Michod, R. E. and Levin, B. R., eds. 1988. *The Evolution of Sex: An Examination of Current Ideas.* Sinauer, New York.

Otto, S. P., and Nuismer, S. L. 2004. Species interactions and the evolution of sex. *Science* 304: 1018–1020.

Ridley, M. 1994. *The Red Queen.* Macmillan, New York.

Stearns, S. C., ed. 1987. *The Evolution of Sex and Its Consequences.* Birkhauser Verlag Basel.

West, S. A., Lively, C. M., and Read, A. F. 1999. A pluralist approach to sex and recombination. *Journal of Evolutionary Biology* 12: 1003–1012.

Williams, G. C. 1975. *Sex and Evolution.* Princeton University Press, Princeton.

CHAPTER 5: WHEN SEX MAKES YOU SICK

Altizer, S., Nunn, C. L., Thrall, P. H., Gittleman, J. L., Antonovics, J., Cunningham, A. A., Dobson, A. P., Ezenwa, V., Jones, K. E., Pedersen, A. B., Poss, M., and Pulliam, J. R. C. 2003. Social organization and parasite risk in mammals: Integrating theory and empirical studies. *Annual Reviews of Ecology, Evolution and Systematics* 34: 517–547.

Boots, M., and Knell, R. 2002. The evolution of risky behaviour in the presence of a sexually transmitted disease. *Proceedings of the Royal Society of London B:* 269: 585–589.

Briskie, J. V., and Montgomerie, R. 2001. Efficient copulation and the evolutionary loss of the avian intromittent organ. *Journal of Avian Biology* 32: 184–187.

Caldwell, T. 2002. *Shivers.* www.sensesofcinema.com/contents/01/19/cteq/shivers.html.

Centers for Disease Control and Prevention. Fact Sheet: Syphilis. www.cdc.gov/std/Syphilis/STDFact-Syphilis.htm.

Hurst, G. D. D., Webberley, K. M., and Knell, R. 2006. The role of parasites of insect reproduction in the diversification of insect reproductive processes. *Insect Evolutionary Ecology (Proceedings of the Royal Entomological Society's 22nd Symposium),* M. D. E. Fellowes, G. J. Holloway, and J. Rolff , eds. Oxford University Press, Oxford.

Knell, R. J. 2004. Syphilis in Renaissance Europe: Rapid evolution of an introduced sexually transmitted disease? *Proceedings of the Royal Society of London B* (Suppl.) 271: S174–S176.

Kokko, H., Ranta, E., Ruxton, G., and Lundberg, P. 2002. Sexually transmitted disease and the evolution of mating systems. *Evolution* 56: 1091–1100.

Levin, B. R., Bull, J. J., and Stewart, F. M. 2001. Epidemiology, evolution, and future of the HIV/AIDS pandemic. *Emerging Infectious Diseases* 7 (Suppl.): 505–511.

Lockhart, A. B., Thrall, P. H., and Antonovics, J. 1996. Sexually transmitted diseases in animals: Ecological and evolutionary implications. *Biological Reviews* 71: 415–471.

Lombardo, M. P. 1998. On the evolution of sexually transmitted diseases in birds. *Journal of Avian Biology* 29: 314–321.

Lombardo, M. P., Thorpe, P. A., and Power, H. W. 1999. The beneficial sexually transmitted microbe hypothesis of avian copulation. *Behavioral Ecology* 10: 333–350.

Nunn, C. L., 2003. Behavioural defences against sexually transmitted diseases in primates. *Animal Behaviour* 66: 37–48.

Nunn, C. L., and Altizer, S. M. 2004. Sexual selection, behaviour and sexually transmitted diseases. *Sexual Selection in Primates: New and Comparative Perspectives.* P. M. Kappeler and C. P. van Schaik, eds. Cambridge University Press, Cambridge.

Nunn, C. L., Gittleman, J. L., and Antonovics, J. 2000. Promiscuity and the primate immune system. *Science* 290: 1168–1170.

Sheldon, B. C. 1993. Sexually transmitted disease in birds: Occurrence and evolutionary significance. *Philosophical Transactions of the Royal Society of London B* 339: 491–497.

Thrall, P. H., and Antonovics, J. 1997. Polymorphism in sexual versus non-sexual disease transmission. *Proceedings of the Royal Society of London B* 264: 581–587.

Thrall, P. H., Antonovics, J., and Dobson, A. P. 2000. Sexually transmitted diseases in polygynous mating systems: Prevalence and impact on reproductive success. *Proceedings of the Royal Society of London B* 267: 1555–1563.

Thrall, P. H., Antonovics, J., and Wilson, W. G. 1998. Allocation to sexual versus nonsexual disease transmission. *American Naturalist* 151: 29–45.

Timms, P. 1998. Chlamydia—modern day scourge of mice and men. *Microbiology Australia,* September. 22–25.

Webberley, K. M., Hurst, G. D. D., Husband, R. W., Schulenburg, J. H. G. V. D., Sloggett, J. J., Isham, V., Buszko, J., and Majerus, M. E. N. 2004. Host reproduction and a sexually transmitted disease: Causes and consequences of *Coccipolipus hippodamiae* distribution on coccinellid beetles. *Journal of Animal Ecology* 73: 1–10.

Wellcome News. 2002. Playing the field. Q1: 22.

CHAPTER 6: **THE SICKER SEX**

Ananthaswamy, A. 2001. The inner strength that keeps women going. NewScientist.com, September 12.

Kinsella, K., and Gist, Y. J. 1998. Mortality and health. International Brief: Gender and Aging, U.S. Department of Commerce, Economics and Statistics Administration, Bureau of the Census.

Kraaijeveld, K., Kraaijeveld-Smit, F. J. L., and Adcock, G. J. 2003. Does female mortality drive male semelparity in dasyurid marsupials? *Proceedings of the Royal Society of London B* (Suppl.) 270: S252–S253.

Kreeger, K. Y. 2002. Sex-based longevity. *The Scientist* 16: 34, May 13.

Kruger, D. J., and Nesse, R. M. 2004. Sexual selection and the male:female mortality ratio. *Evolutionary Psychology* 2: 66–85.

Liker, A., and Szekely, T. 2005. Mortality costs of sexual selection and parental care in natural populations of birds. *Evolution* 59: 890–897.

Liu, P. Y., Death, A. K., and Handelsman, D. J. 2003. Androgens and cardiovascular disease. *Endocrine Reviews* 24: 313–340.

Moore, S. L., and Wilson, K. 2002. Parasites as a viability cost of sexual selection in natural populations of mammals. *Science* 297: 2015–2018.

Morales-Montor, J., Chavarria, A., De Leon, M. A., Del Castillo, L .I., Escobedo, E. G., Sanchez, E. N., Vargas, J. A., Hernandez-Flores, M., Romo-Gonzalez, T., and Larralde, C. 2004. Host gender in parasitic infections of mammals: An evaluation of the female host supremacy paradigm. *Journal of Parasitology* 90: 531–546.

Oakwood, M., Bradley, A. J., and Cockburn, A. 2001. Semelparity in a large marsupial. *Proceedings of the Royal Society of London B* 268: 407–411.

Owens, I. P. F. 2002. Sex differences in mortality rate. *Science* 297: 2008–2009.

Shettles, L. B. 1958. Biological sex differences with special reference to disease, resistance and longevity. *The Journal of Obstetrics and Gynaecology of the British Empire* LXV: 288–295.

Skorping, A., and Jensen, K. H. 2004. Disease dynamics: All caused by males? *Trends in Ecology and Evolution* 19: 219–220.

Teriokhin, A. T., Budilova, E. V., Thomas, F., and Guegan, J.-F. 2004. Worldwide variation in life-span sexual dimorphism and sex-specific environmental mortality rates. *Human Biology* 76: 623–641.

Wiklund, C., Gorrhard, K., and Nylin, S. 2003. Mating system and the evolution of sex-specific mortality rates in two nymphalid butterflies. *Proceedings of the Royal Society of London B* 270: 1823–1828.

Woo, J., and Ho, S. 2003. Sex differences in life expectancy. *XX vs. XY* 1: 91–93.

Worldbank.org report: Beyond Economic Growth. Chapter 8: Health and Longevity.

World Health Organization. 1998. Gender and health: technical paper. WHO/FRH/WHD/98.16.

Zuk, M. 1990. Reproductive strategies and sex differences in disease susceptibility: An evolutionary viewpoint. *Parasitology Today* 6: 231–233.

Zuk, M., and McKean, K. A. 1996. Sex differences in parasite infections: Patterns and processes. *International Journal for Parasitology* 26: 1009–1024.

CHAPTER 7: **PARASITES AND PICKING THE PERFECT PARTNER**

Andersson, M. 1994. *Sexual Selection*. Princeton University Press, Princeton.

Burnham, C., Chapman, J. F., Gray, P. B., McIntyre, M. H., Lipson, S. F., and Ellison, P. T. 2003. Men in committed, romantic relationships have lower testosterone. *Hormones and Behavior* 44: 119–122.

Clotfelter, E. D., O'Neal, D. M., Gaudioso, G. M., Casto, J. M., Parker-Renga, I. M., Snajdr, E. A., Duffy, D. L., Nolan, V., and Ketterson, E. D. 2004. Consequences of elevating plasma testosterone in females of a socially monogamous songbird: Evidence of constraints on male evolution? *Hormones and Behavior* 46: 171–178.

Duffy, D. L., and Ball, G. F. 2002. Song predicts immunocompetence in male European starlings (*Sturnus vulgaris*). *Proceedings of the Royal Society of London B* 269: 847–852.

Foerster, K., and Kempenaers, B. 2004. Experimentally elevated plasma levels of testosterone do not increase male reproductive success in blue tits. *Behavioral Ecology and Sociobiology* 56: 482–490.

Folstad, I., and Karter, A. J. 1992. Parasites, bright males and the immunocompetence handicap. *The American Naturalist* 139: 603–622.

Hamilton, W. D., and Zuk, M. 1982. Heritable true fitness and bright birds: A role for parasites? *Science* 213: 384–387.

Hamilton, W. J., and Poulin, R. 1997. The Hamilton and Zuk hypothesis revisited: A meta-analytical approach. *Behaviour* 134: 299–320.

Hanssen, S. A., Hasselquist, D., Folstad, I., and Erikstad, K. E. 2004. Costs of immunity: Immune responsiveness reduces survival in a vertebrate. *Proceedings of the Royal Society of London B* 271: 925–930.

Johns, J. L. 1997. The Hamilton-Zuk theory and initial test: An examination of some parasitological criticisms. *International Journal for Parasitology* 27: 1269–1288.

Kilpimaa, J., Alatalo R. V., and Siitari, H. 2004. Trade-offs between sexual advertisement and immune function in the pied flycatcher (*Ficedula hypoleuca*). *Proceedings of the Royal Society of London B* 271: 245–250.

Klein, S. L. 2000. The effects of hormones on sex differences in infection: From genes to behavior. *Neuroscience and Biobehavioral Reviews* 24: 627–638.

Liljedal, S., Flostad, I., and Skarstein, F. 1999. Secondary sex traits, parasites, immunity and ejaculate quality in the Arctic charr. *Proceedings of the Royal Society of London B* 266: 1893–1898.

Lindström, K., and Lundström, J. 2000. Male greenfinches (*Carduelis chloris*) with brighter ornaments have higher virus infection clearance rate. *Behavioral Ecology and Sociobiology* 48: 44–51.

Low, B. S. 1990. Marriage systems and pathogen stress in human societies. *American Zoologist* 30 (2): 325–339.

Måsvær, M., Liljedal, S., and Folstad, I. 2004. Are secondary sex traits, parasites and immunity related to variation in primary sex traits in the Arctic charr? *Proceedings of the Royal Society of London B* (Suppl.) 271: S40–S42.

McLennan, D. A., and Brooks, D. R. 1991. Parasites and sexual selection—a macroevolutionary perspective. *Quarterly Review of Biology* 66: 255–286.

Moore, S. L., and Wilson, K. 2002. Parasites as a viability cost of sexual selection in natural populations of mammals. *Science* 297: 2015–2018.

Mougeot, F., and Redpath, S. M. 2004. Sexual ornamentation relates to immune function in male red grouse. *Journal of Avian Biology* 35: 425–433.

Møller, A. P. 1990. Effects of a hematophagous mite on the barn swallow (*Hirundo rustica*)—a test of the Hamilton and Zuk hypothesis. *Evolution* 44 (4): 771–784.

Møller, A. P., Christe, P., and Lux, E. 1999. Parasitism, host immune function, and sexual selection. *The Quarterly Review of Biology* 74: 3–20.

Pagel, M., and Bodmer, W. 2003. A naked ape would have fewer parasites. *Proceedings of the Royal Society of London B* 270: S117–S119.

Reinhardt, K., Naylor, R., and Siva-Jothy, M. T. 2003. Reducing a cost of traumatic insemination: Female bedbugs evolve a unique organ. *Proceedings of the Royal Society of London B* 270: 2371–2375.

Roberts, M. L., Buchanan, K. L., and Evans, M. R. 2004. Testing the immunocompetence handicap hypothesis: A review of the evidence. *Animal Behaviour* 68: 227–239.

Rolff, J., and Siva-Jothy, M. T. 2002. Copulation corrupts immunity: A mechanism for a cost of mating in insects. *Proceedings of the National Academy of Sciences, U.S.A.* 99: 9916–9918.

Skarstein, F., Folstad, I., and Liljedal, S. 2001. Whether to reproduce or not: Immune suppression and costs of parasites during reproduction in the Arctic charr. *Canadian Journal of Zoology* 79: 271–278.

Skau, P. A., and Folstad, I. 2003. Do bacterial infections cause reduced ejaculate quality? A meta-analysis of antibiotic treatment of male infertility. *Behavioral Ecology* 14: 40–47.

Stutt, A. D., and Siva-Jothy, M. T. 2001. Traumatic insemination and sexual conflict in the bedbug *Cimex lectularius*. *Proceedings of the National Academy of Sciences, U.S.A.* 98: 5863–5867.

Wedekind, C., and Folstad, I. 1994. Adaptive or nonadaptive immunosuppression by sex hormones? *The American Naturalist* 143: 936–938.

Zuk, M. 1984. A charming resistance to parasites. *Natural History* 93: 28–34.

Zuk, M. 1990. Reproductive strategies and disease susceptibility: An evolutionary viewpoint. *Parasitology Today* 6: 231–233.

Zuk, M. 1992. The role of parasites in sexual selection: Current evidence and future directions. *Advances in the Study of Behavior* 21: 39–68.

Zuk, M., and Johnsen, T. S. 2000. Social environment and immunity in male red jungle fowl. *Behavioral Ecology* 11: 146–153.

Zuk, M., Thornhill, R., Johnson, K., and Ligon, J. D. 1990. Parasites and mate choice in red jungle fowl. *American Zoologist* 30: 235–244.

CHAPTER 8: **WHEN HOW YOU FEEL IS HOW YOU LOOK**

Alonso-Alvarez, C., Bertrand, S., Devevey, G., Gaillard, M., Prost, J., Faivre, B., and Sorci, G. 2004. An experimental test of the dose-dependent effect of carotenoids and immune activation on sexual signals and antioxidant activity. *American Naturalist* 164: 651–659.

Bloom, D. F. 2004. Is acne really a disease?: A theory of acne as an evolutionarily significant, high-order psychoneuroimmune interaction timed to cortical development with a crucial role in mate choice. *Medical Hypotheses* 62: 462–469.

Blount, J. D. 2004. Carotenoids and life-history evolution in animals. *Archives of Biochemistry and Biophysics* 430: 10–15.

Blount, J. D., Metcalfe, N. B., Arnold, K. E., Surai, P .F., Devevey, G. L., and Monaghan, P. 2003. Neonatal nutrition, adult antioxidant defences and sexual attractiveness in the zebra finch. *Proceedings of the Royal Society of London B* 270: 1691–1696.

Blount, J. D., Metcalfe, N. B., Birkhead, T. R., and Surai, P. F. 2003. Carotenoid modulation of immune function and sexual attractiveness in zebra finches. *Science* 300: 125–127.

Boelsma, E., Hendriks, H. F. J., and Roza, L. 2001. Nutritional skin care: Health effects of micronutrients and fatty acids. *American Journal of Clinical Nutrition* 73: 853–864.

Cellerino, A. 2002. Facial attractiveness and species recognition: An elementary deduction? *Ethology, Ecology & Evolution* 14: 227–237.

Chew, B. P., and Park, J. S. 2004. Carotenoid action on the immune response. *Journal of Nutrition* 134: 257S–261S

Faivre, B., Preault, M., Salvadori, F., Thery, M., Gaillard, M., and Cezilly, F. 2003. Bill colour and immunocompetence in the European blackbird. *Animal Behaviour* 65: 1125–1131.

Faivre, B., Gregoire, A., Preault, M., and Cezilly, F. 2003. Immune activation rapidly mirrored in a secondary sexual trait. *Science* 300: 103.

Fink, B., Grammer, K., and Thornhill, R. 2001. Human (*Homo sapiens*) facial attractiveness in relation to skin texture and color. *Journal of Comparative Psychology* 115: 92–99.

Fink, B., and Penton-Voak, I. 2002. Evolutionary psychology of facial attractiveness. *Current Directions in Psychological Science* 11: 154–158.

Good, M. 2000. Vaccination: The facts, the fears, the future (Book Review). *Immunology and Cell Biology* 78: 649.

Grammer, K., Fink, B., Moller, A. P., and Thornhill, R. 2003. Darwinian aesthetics: Sexual selection and the biology of beauty. *Biological Reviews* 78: 385–407.

Grether, G. F., Hudon, J., and Endler, J. A. 2001. Carotenoid scarcity, synthetic pteridine pigments and the evolution of sexual coloration in guppies (*Poecilia reticulata*). *Proceedings of the Royal Society of London B* 268: 1245–1253.

Gurel, M. S., Ulukanligil, M., and Ozbilge, H. 2002. Cutaneous leishmaniasis in Sanliurfa: Epidemiologic and clinical features of the last four years (1997–2000). *International Journal of Dermatology* 41: 32–37.

Hill, G. E. 1999. Is there an immunological cost to carotenoid-based ornamental coloration? *American Naturalist* 154: 589–595.

Hill, G. E., Inouye, C. Y., and Montgomerie, R. 2002. Dietary carotenoids predict plumage coloration in wild house finches. *Proceedings of the Royal Society of London B* 269: 1119–1124.

Hinsz, V. B., Matz, D. C., and Patience, R. A. 2001. Does women's hair signal reproductive potential? *Journal of Experimental Social Psychology* 37: 166–172.

Hughes, D. A. 1999. Effects of carotenoids on human immune function. *Proceedings of the Nutrition Society* 58: 713–718.

Jones, B. C., Little, A. C., Burt, D. M., and Perrett, D. I. 2004. When facial attractiveness is only skin deep. *Perception* 33: 569–576.

Kalick, S. M., Zebrowitz, L. A, Langlois, J. H., and Johnson, R. M. 1998. Does human facial attractiveness honestly advertise health? Longitudinal data on an evolutionary question. *Psychological Science* 9: 8–13.

Kellett, S., and Gilbert, P. 2001. Acne: A biopsychosocial and evolutionary perspective with a focus on shame. *British Journal of Health Psychology* 6: 1–24.

Kerr, J. 1960. *The Snake Has All the Lines.* Doubleday & Co. New York.

Long, K. Z., and Nanthakumar, N. 2004. Energetic and nutritional regulation of the adaptive immune response and trade-offs in ecological immunology. *American Journal of Human Biology* 16: 499–507.

Lozano, G. A. 1994. Carotenoids, parasites, and sexual selection. *Oikos* 70: 309–311.

McGraw, K. J., and Hill, G. E. 2000. Differential effects of endoparasitism on the expression of carotenoid- and melanin-based ornamental coloration. *Proceedings of the Royal Society of London B* 267: 1525–1531.

McGraw, K. J., and Ardia, D. R. 2003. Carotenoids, immunocompetence, and the information content of sexual colors: An experimental test. *American Naturalist* 162: 704–712.

McGraw, K. J., Wakamatsu, K., Clark, A. B., and Yasukawa, K. 2004. Red-winged blackbirds *Agelaius phoeniceus* use carotenoid and melanin pigments to color their epaulets. *Journal of Avian Biology* 35: 543–550.

McGraw, K. J. 2004. Colorful songbirds metabolize carotenoids at the integument. *Journal of Avian Biology* 35: 471–476.

Mesko, N., and Bereczkei, T. 2004. Hairstyle as an adaptive means of displaying phenotypic quality. *Human Nature* 15: 251–270.

Negro, J. J., Grande, J. M., Tella, J. L., Garrido, J., Hornero, D., Donazar, J. A., Sanchez-Zapata, J. A., Benitez, J. R., and Barcell, M. An unusual source of essential carotenoids. *Nature* 416: 807–808.

Negro, J. J., Margalida, A., Hiraldo, F., and Heredia, R. 1999. The function of cosmetic coloration of bearded vultures: When art imitates life. *Animal Behaviour* 58: F14–F17.

Negro, J. J., Margalida, A., Torres, M. J., Grande, J. M., Hiraldo, F., and Heredia, R. 2002. Iron oxides in the plumage of bearded vultures: Medicine or cosmetics? *Animal Behaviour* 64: F3–F7.

Olson, V. A., and Owens, I. P. F. 1998. Costly sexual signals: Are carotenoids rare, risky or required? *Trends in Ecology and Evolution* 13: 510–514.

Peters, A., Denk, A. G., Delhey, K., and Kempenaers, B. 2004. Carotenoid-based bill colour as an indicator of immunocompetence and sperm performance in male mallards. *Journal of Evolutionary Biology* 17: 1111–1120.

Peters, A., Delhey, K., Denk, A. G., and Kempenaers, B. 2004. Trade-offs between immune investment and sexual signaling in male mallards. *American Naturalist* 164: 51–59.

Rhodes, G., Chan, J., Zebrowitz, L. A., and Simmons, L. W. 2003. Does sexual dimorphism in human faces signal health? *Proceedings of the Royal Society of London B* (Suppl.) 270: S93–S95.

Saino, N., Ninni, P., Calza, S., Martinelli, R., De Bernardi, F., and Møller, A. P. 2000. Better red than dead: Carotenoid-based mouth coloration reveals infection in barn swallow nestlings. *Proceedings of the Royal Society of London B* 267: 57–61.

Tella, J. L., Figuerola, J., Negro, J. J., Blanco, G., Rodriguez-Estrella, R., Forero, M. G., Blazquez, M. C., Green, A. J., and Hiraldo, F. 2004. Ecological, morphological and phylogenetic correlates of interspecific variation in plasma carotenoid concentration in birds. *Journal of Evolutionary Biology* 17: 156–164.

Zuk, M., and Johnsen, T. S. 1998. Seasonal changes in the relationship between ornamentation and immune response in red jungle fowl. *Proceedings of the Royal Society of London B* 265: 1631–1635.

CHAPTER 9: **TAKING CARE**

Billing, J., and Sherman, P. W. 1998. Antimicrobial functions of spices: Why some like it hot. *The Quarterly Review of Biology* 73: 3–49.

Biser, J. A. 1998. Really wild remedies—medicinal plant use by animals. *ZooGoer* 27 (1).

Callahan, G. N. 2003. Eating dirt. *Emerging Infectious Diseases* 9: 1016–1021.

Carrai, V., Borgognini-Tarli, S. M., Huffman, M. A., and Bardi, M. 2003. Increase in tannin consumption by sifaka (*Propithecus verreauxi verreauxi*) females during the birth season: A case for self-medication in prosimians? *Primates* 44: 61–66.

Carroll, J. F., Kramer, M., Weldon, P. J., and Robbins, R. G. 2005. Anointing chemicals and ectoparasites: Effects of benzoquinones from millipedes on the lone star tick. *Journal of Chemical Ecology* 31: 63–75.

Christe, P., Oppliger, A., Bancala, F., Castella, G., and Chapuisat, M. 2003. Evidence for collective medication in ants. *Ecology Letters* 6: 19–22.

Cousins, D., and Huffman, M. A. 2002. Medicinal properties in the diet of gorillas: An ethno-pharmacological evaluation. *African Study Monographs* 23: 65–89.

Curtis, V., and Jolly, A. 2003. Sick as a parrot. *London Review of Books* 25, July 10.

Curtis, V., Aunger, R., and Rabie, T. 2004. Evidence that disgust evolved to protect from risk of disease. *Proceedings of the Royal Society of London B* (Suppl.) 271: S131–S133.

DeJoseph, M., Taylor, R. S. L., Baker, M., and Aregullin, M. 2002. Fur-rubbing behavior of capuchin monkeys. *Journal of the American Academy of Dermatology* 46: 924–925.

Diamond, J. M. 1999. Dirty eating for healthy living. *Nature* 400: 120–121.

Dominy, N. J., Davoust, E., and Minekus, M. 2004. Adaptive function of soil consumption: An *in vitro* study modeling the human stomach and small intestine. *Journal of Experimental Biology* 207: 319–324.

Eckstein, R. A., and Hart, B. L. 2000. Grooming and control of fleas in cats. *Applied Animal Behaviour Science* 68: 141–150.

Engel, C. 2002. *Wild Health: Lessons in Natural Wellness from the Animal Kingdom.* Houghton Mifflin, New York.

Faulkner, J., Schaller, M., Park, J. H., and Duncan, L. A. 2004. Evolved disease-avoidance mechanisms and contemporary xenophobic attitudes. *Group Processes & Intergroup Relations* 7: 333–353.

Freeland, W. J. 1976. Pathogens and the evolution of primate sociality. *BioTropica* 8: 12–24.

Furlow, B. 2000. Kills all known germs. *New Scientist,* January 22.

Gehlbach, F. R., and Baldridge, R. S. 1987. Live blind snakes (*Leptotyphlops dulcis*) in eastern screech owl (*Otus asio*) nests: A novel commensalism. *Oecologia* 71: 560–563.

Geissler, P. W. 2000. The significance of earth-eating: Social and cultural aspects of geophagy among Luo children. *Africa* 70: 653–682.

Geissler, P. W., Mwaniki, D., Thiong, F., and Friis, H. 1998. Geophagy as a risk factor for geohelminth infections: A longitudinal study of Kenyan primary schoolchildren. *Transactions of the Royal Society of Tropical Medicine and Hygiene* 92: 7–11.

Hart, B. L., Korinek, E., and Brennan, P. 1987. Postcopulatory genital grooming in male rats: Prevention of sexually transmitted infections. *Physiology and Behavior* 41: 321–325.

Hart, B. L. 1988. Biological basis of the behavior of sick animals. *Neuroscience and Biobehavioral Reviews* 12: 123–127.

Hart, B. L. 1990. Behavioral adaptations to pathogens and parasites: Five strategies. *Neuroscience and Biobehavioral Reviews* 14: 273–294.

Hart, B. L. 1992. Behavioral adaptations to parasites: An ethological approach. *Journal of Parasitology* 78: 256–265.

Hart, B. L. 1994. Behavioural defense against parasites: Interaction with parasite invasiveness. *Parasitology* 109: S139–S151.

Hart, B. L. 2000. Role of grooming in biological control of ticks. *Annals of the New York Academy of Sciences* 916: 565–569.

Hemmes, R.B., Alvarado, A., and Hart, B.L. 2002. Use of California bay foliage by wood rats for possible fumigation of nest-borne ectoparasites. *Behavioral Ecology* 13: 381–385.

Huffman, M. A. 2001. Self-medicative behavior in the African great apes: An evolutionary perspective into the origins of human traditional medicine. *BioScience* 51: 651–661.

Huffman, M. A. 1997. Current evidence for self-medication in primates: A multidisciplinary perspective. *Yearbook of Physical Anthropology* 40: 171–200.

Huffman, M. A. 2003. Animal self-medication and ethno-medicine: Exploration and exploitation of the medicinal properties of plants. *Proceedings of the Nutrition Society* 62: 371–381.

Huffman, M. A., and Caton, J. M. 2001. Self-induced increase of gut motility and the control of parasitic infections in wild chimpanzees. *International Journal of Primatology* 22: 329–346.

Huffman, M. A., and Hirata, S. 2004. An experimental study of leaf swallowing in captive chimpanzees: Insights into the origin of a self-medicative behavior and the role of social learning. *Primates* 45: 113–118.

Karban, R., and English-Loeb, G. Tachinid parasitoids affect host plant choice by caterpillars to increase caterpillar survival. *Ecology* 78: 603–611.

Knezevich, M. 1998. Geophagy as a therapeutic mediator of endoparasitism in a free-ranging group of rhesus macaques (*Macaca mulatta*). *American Journal of Primatology* 44: 71–82.

Krishnamani, R., and Mahaney, W. C. 2000. Geophagy among primates: Adaptive significance and ecological consequences. *Animal Behaviour* 59: 899–915.

Lozano, G. A. 1998. Parasitic stress and self-medication in wild animals. *Advances in the Study of Behavior* 27: 291–317.

Luoba, A. I., Geissler, P. W., Estambale, B., Ouma, J. H., Alusala, D., Ayah, R., Mwaniki, D., Magnussen, P., and Friis, H. 2005. Earth-eating and reinfection with intestinal helminthes among pregnant and lactating women in western Kenya. *Tropical Medicine and International Health* 10: 220–227.

Mahaney, W. C., and Krishnamani, R. 2003. Understanding geophagy in animals: Standard procedures for sampling soils. *Journal of Chemical Ecology* 29: 1503–1523.

Mooring, M. S., and Hart, B. L. 1992. Animal grouping for protection from parasites: Selfish herd and encounter-dilution effects. *Behaviour* 123: 173–193.

Mooring, M. S., McKenzie, A. A., and Hart, B. L. 1996. Role of sex and breeding status in grooming and total tick load of impala. *Behavioral Ecology and Sociobiology* 39: 259–266.

Mooring, M. S., Blumstein, D. T., and Stoner, C. J. 2004. The evolution of parasite-defence grooming in ungulates. *Biological Journal of the Linnean Society* 81: 17–37.

Petit, C., Hossaert-McKey, M., Perret, P., Blondel, J., and Lambrechts, M. M. 2002. Blue tits use selected plants and olfaction to maintain an aromatic environment for nestlings. *Ecology Letters* 5: 585–589.

Revis, H., and Waller, D. A. 2004. Bactericidal and fungicidal activity of ant chemicals on feather parasites: An evaluation of anting behavior as a method of self-medication in songbirds. *The Auk* 121: 1262–1268.

Sallan, K. I., Ishioroshi, M., and Samejima, K. 2004. Antioxidant and antimicrobial effects of garlic in chicken sausage. *Lebensm.-Wiss. U.-Technol.* 37: 849–855.

Sapolsky, R. M. 1994. Fallible instinct: A dose of skepticism about the medicinal "knowledge" of animals. *The Sciences* 34: 1–4.

Sherman, P. W., and Billing, J. 1999. Darwinian gastronomy: Why we use spices. *BioScience* 49: 453–463.

Sherman, P. W., and Flaxman, S. M. 2001. Protecting ourselves from food. *American Scientist* 89: 142–151.

Sherman, P. W., and Hash, G. A. 2001. Why vegetable recipes are not very spicy. *Evolution and Human Behavior* 22: 147–163.

Villalba, J. J., Provenza, F. D., and Shaw, R. 2006. Sheep self-medicate when challenged with illness-inducing foods. *Animal Behaviour* 71: 1131–1139.

Vtazkova, S. K., Long, E., Paul, A., and Glendinning, J. I. 2001. Mice suppress malaria infection by sampling a "bitter" chemotherapy agent. *Animal Behaviour* 61: 887–894.

Weldon, P. J. 2004. Defensive anointing: Extended chemical phenotype and unorthodox ecology. *Chemoecology* 14: 1–4.

Weldon, P. J., Aldrich, J. R., Klun, J. A., Oliver, J. E., and Debboun, M. 2003. Benzoquinones from millipedes deter mosquitoes and elicit self-anointing in capuchin monkeys (*Cebus* spp.). *Naturwissenschaften* 90: 301–304.

CHAPTER 10: BAD, BUT NOT WEIRD: THE REAL EMERGING DISEASES

Bakalar, N. 2005. More diseases pinned on old culprit: Germs. *New York Times,* May 17.

Belkin, L. 2005. Can you catch obsessive-compulsive disorder? *New York Times,* May 22.

Brooks, D. R., McLennan, D. A., and Leon-Regagnon, V. Unpublished abstract. Ecological fitting and evolutionary accidents waiting to happen: Establishment and rapid spread of an introduced pathogen.

Brooks, D. R., and Ferrao, A. 2005. The historical biogeography of co-evolution: Emerging infectious diseases are evolutionary accidents waiting to happen. *Journal of Biogeography* 32: 1291–1299.

Brooks, D. R., and Hoberg, E. P. Unpublished manuscript. Systematics and emerging infectious diseases: From management to solution.

Brown, A. S., Schaefer, C. A., Quesenberry, C. P., Liu, L. Y., Babulas, V. P., and Susser, E. S. 2005. Maternal exposure to toxoplasmosis and risk of schizophrenia in adult offspring. *American Journal of Psychiatry* 162: 767–773.

Brown, W. A. 2005. PANDAS: Nonexistent or simply rare? *Applied Neurology,* June 1. www.appneurology.com.

Calvert, S., and Kohn, D. 2005. Baffling diseases emerging from Africa. *Seattle Times,* May 16.

Carbone, K. M., Luftig, R. B., and Buckley, M. R. 2005. Microbial triggers of chronic human illness. *American Academy of Microbiology,* Washington DC.

Cochran, G. M., Ewald, P. W., and Cochran, K. D. 2000. Infectious causation of disease: An evolutionary perspective. *Perspectives in Biology and Medicine* 43: 406–448.

Cunningham, A .A., Daszak, P., and Rodriguez, J. P. 2003. Pathogen pollution: Defining a parasitological threat to biodiversity conservation. *Journal of Parasitology* 89 (Suppl.): S78–S83.

Daszak, P., Tabor, G. M., Kilpatrick, A. M., Epstein, J., and Plowright, R. 2004. Conservation medicine and a new agenda for emerging diseases. *Annals of the New York Academy of Sciencies* 1026: 1–11.

Denver Naturopathic Clinic News. 2005. MS update: Siblings, Epstein-Barr virus and other links. www.denvernaturopathic.com/news/ebvMS.html.

Dinello, D. 2001. Virus horror! Salon.com, August 9.

Elston, D. M. 2005. New and emerging infectious diseases. *Journal of the American Academy of Dermatology* 52: 1062–1068.

Epstein, P., Chivian, E., and Frith, K. 2003. Emerging diseases threaten conservation. *Environmental Health Perspectives* 111: A506–A507.

Ewald, P. 2002. *Plague Time: The New Germ Theory of Disease.* Anchor, New York.

Fauci, A. S., Touchette, N. A., and Folkers, G. K. 2005. Emerging infectious diseases: A 10-year perspective from the National Institute of Allergy and Infectious Diseases. *Emerging Infectious Diseases* 11: 519–525.

Finch, C. E., and Crimmins, E. M. 2004. Inflammatory exposure and historical changes in human life-spans. *Science* 305: 1736–1739.

Gibbs, M. J., and Gibbs, A. J. 2006. Was the 1918 pandemic caused by a bird flu? *Nature* 440: E8–E10.

Ginsburg, J. 2003. Diseases of the mind. *Newsweek International,* Dec. 1.

Gluckman, P. D., and Hanson, M. A. 2004. Living with the past: Evolution, development, and patterns of disease. *Science* 305: 1733–1736.

Gozlan, R. E., St-Hilaire, S., Feist, S. W., Martin, P., and Kent, M. L. 2005. Disease threat to European fish. *Nature* 435: 1046.

Grady, D., and Kolata, G. 2006. How serious is the risk of avian flu? *New York Times,* March 27.

Guernier, V., Hochberg, M. E., and Guegan, J. F. O. 2004. Ecology drives the worldwide distribution of human diseases. *PloS Biology* 2: 740–746.

Guterl, F. 2003. The battle against bugs gets serious. *Newsweek,* December 29.

Hawaleshka, D. 2005. Bracing for bird flu. *MacLean's,* March 16.

Hobson, K. 2003. Pet problems. USNews.com, June 23.

Hooper, J. 1999. A new germ theory. *The Atlantic* online. February.

Kaiser, J. 2005. More infectious diseases emerge in north. *Science* 307: 1190.

Kilpatrick, A. M., Kramer, L. D., Campbell, S. R., Alleyne, E. O., Dobson, A. P., and Daszak, P. West Nile virus risk assessment and the bridge vector paradigm. *Emerging Infectious Diseases* 11: 425–429.

Kinoshita, J. 2004. Pathogens as a cause of Alzheimer's disease. *Neurobiology of Aging* 25: 639–640.

Klar, A. J. S. 2004. A genetic mechanism implicates chromosome 11 in schizophrenia and bipolar diseases. *Genetics* 167: 1833–1840.

Kutz, S. J., Hoberg, E. P., Nagy, J., Polley, L., and Elkin, B. 2004. "Emerging" parasitic infections in arctic ungulates. *Integrative and Comparative Biology* 44: 109–118.

Lax, A. J. 2005. Bacterial toxins and cancer—a case to answer? *Nature Reviews Microbiology* 3: 343–349.

Lederberg, J. 1997. Infectious disease as an evolutionary paradigm. *Emerging Infectious Diseases* 3: 417–423.

Ledgerwood, L. G., Ewald, P. W., and Cochran, G. M. 2003. Genes, germs, and schizophrenia—an evolutionary perspective. *Perspectives in Biology and Medicine* 46: 317–348.

Lipkin, W. I., Hornig, M. 2004. Psychotropic viruses. *Current Opinion in Microbiology* 7: 420–425.

Lyons, S. K., Smith, F. A., Wagner, P. J., White, E.P., and Brown, J. H. 2004. Was a "hyperdisease" responsible for the late Pleistocene megafaunal extinction? *Ecology Letters* 7: 859–868.

Maugh, Thomas II. 1999. Spreading a new idea on disease: Mounting evidence may link viruses and bacteria to everything from gallstones to Alzheimer's. *Los Angeles Times.* April 22.

McMichael, T. 2002. Fine battlefield reporting, but it's time to stop the war metaphor (review of *Secret Agents: The Menace of Emerging Infections,* by Madeline Drexler). *Science* 295: 1469.

Mirsky, S. 2001. A host with infectious ideas. *Scientific American,* May.

Morens, D. M., Folkers, G. K., and Fauci, A. S., 2004. The challenge of emerging and re-emerging infectious diseases. *Nature* 430: 242–249.

Murphy, F. A., 1998. Emerging zoonoses. *Emerging Infectious Diseases* 4: 429–435.

Normile, D. 2006. Avian influenza—studies suggest why few humans catch the H5N1 virus. *Science* 311: 1692.

Parsonnet, J. 2001. The enemy within (review of *Plague Time,* by Paul Ewald). *American Scientist,* May-June.

Pearl, M., and Epstein, J. 2004. Anatomy of an outbreak. January 5, www.sfgate.com.

Ritvo, P., Wilson, K., Willms, D., and Upshur, R. 2005. Vaccines in the public eye. *Nature Medicine Supplement* 11: S20–S24.

Schacter, J. 2001. Review of *Plague Time,* by Paul Ewald. *New England Journal of Medicine* 344: 1175–1176.

Schloegel, L. M., and Daszak, P. Conservation medicine. *Wildlife Tracks* 8: 1–7.

Torchin, M. E., Lafferty, K. D., Dobson, A. P., McKenzie, V. J., and Kuris, A. M., 2003. Introduced species and their missing parasites. *Nature* 421: 628–630.

Wade, N. 2006. Studies suggest avian flu pandemic isn't imminent. *New York Times,* March 23.

Weiss, R. A., and McMichael, A. J., 2004. Social and environmental risk factors in the emergence of infectious diseases. *Nature Medicine Supplement* 10: S70–S76.

Wright, L. 2003. To vanquish a virus. *Scientific American,* July.

Yucesan, C., and Sriram, S. 2001. *Chlamydia pneumoniae* infection of the central nervous system. *Current Opinion in Neurology* 14: 355–359.

CHAPTER 11: **WHO IS IN CHARGE HERE, ANYWAY?**

Adamo, S. A. 2002. Modulating the modulators: Parasites, neuromodulators and host behavioral change. *Brain, Behavior and Evolution* 60: 370–377.

Baudoin, M. 1975. Host castration as a parasitic strategy. *Evolution* 29: 335–352.

Berdoy, M., Webster, J. P., and Macdonald, D. W. 2000. Fatal attraction in rats infected with *Toxoplasma gondii*. *Proceedings of the Royal Society of London B*: 1591–1594.

Brown, S. P. 2005. Do all parasites manipulate their hosts? *Behavioural Processes* 68: 237–240.

Cezilly, F., and Perrot-Minnot, M. J. 2005. Studying adaptive changes in the behaviour of infected hosts: A long and winding road. *Behavioural Processes* 68: 223–228.

Chen, X., Wu, K., and Lun, Z. 2005. Toxoplasmosis researches in China. *Chinese Medical Journal* 118: 1015–1021.

Combes, C. 2005. Manipulations: Variations on the themes of signalling and exaptation. *Behavioural Processes* 68: 211–213.

de Jong-Brink, M., and Koene, J. M. 2005. Parasitic manipulation: Going beyond behaviour. *Behavioural Processes* 68: 229–233.

Dunn, A. M. 2005. Parasitic manipulation of host life history and sexual behaviour. *Behavioural Processes* 68: 255–258.

Eberhard, W. G. 2001. Under the influence: Webs and building behavior of *Plesiometa argyra* (Araneae, Tetragnathidae) when parasitized by *Hymenoepimecis argyraphaga* (Hymenoptera, Ichneumonidae). *Journal of Arachnology* 29: 354–366.

Flegr, J., and Hrdý, I. 1994. Influence of chronic toxoplasmosis on some human personality factors. *Folia Parasitologica* 41: 122–126.

Flegr, J., and Havlíček, J. 1999. Changes in the personality profile of young women with latent toxoplasmosis. *Folia Parasitologica* 46: 22–28.

Flegr, J., Havlíček, J., Kodym, P., Maly, M., and Smahel, Z. 2002. Increased risk of traffic accidents in subjects with latent toxoplasmosis: A retrospective case-control study. *BMC Infectious Diseases* 2: 1–13.

Flegr, J., Preiss, M., Klose, J., Havlíček, J., Vitáková, M., and Kodym, P. 2003. Decreased level of psychobiological factor novelty seeking and lower intelligence in men latently infected with the protozoan parasite *Toxoplasma gondii*. *Biological Psychology* 63: 253–268.

Flegr, J., Zitková, S., Kodym, P., and Frynta, D. 1996. Induction of changes in human behaviour by the parasitic protozoan *Toxoplasma gondii*. *Parasitology* 113: 49–54.

Franz, K., and Kurtz, J. 2002. Altered host behaviour: Manipulation or energy depletion in tapeworm-infected copepods? *Parasitology* 125: 187–196.

Gandon, S. 2005. Parasitic manipulation: A theoretical framework may help. *Behavioural Processes* 68: 247–248.

Gourbal, B. E. F., Righi, M., Petit, G., and Gabrion, C. 2001. Parasite-altered host behavior in the face of a predator: Manipulation or not? *Parasitology Research* 87: 186–192.

Gourbal, B. E. F., Lacroix, A., and Gabrion, C. 2002. Behavioural dominance and *Taenia crassiceps* parasitism in BALB/c male mice. *Parasitology Research* 88: 912–917.

Havlíček, J., Gašová, Z., Smith, A.P., Zvára, K., and Flegr, J. 2001. Decrease of psychomotor performance in subjects with latent "asymptomatic" toxoplasmosis. *Parasitology* 122: 515–520.

Helluy, S., and Holmes, J. C. 2005. Parasitic manipulation: Further considerations. *Behavioural Processes* 68: 205–210.

Hinson, E. R., Shone, S. M., Zink, C., Glass, G. E., and Klein, S. L. 2004. Wounding: The primary mode of Seoul virus transmission among male Norway rats. *American Journal of Tropical Medicine and Hygiene* 70: 310–317.

Hughes, D. P., Katharithamby, J., Turillazzi, S., and Beani, L. 2004. Social wasps desert the colony and aggregate outside if parasitized: Parasite manipulation? *Behavioral Ecology* 15: 1037–1043.

Hughes, D. P. 2005. Parasitic manipulation: A social context. *Behavioural Processes* 68: 263–266.

Hurd, H. 2001. Host fecundity reduction: A strategy for damage limitation? *Trends in Parasitology* 17: 363–368.

Hurd, H. 2003. Manipulation of medically important insect vectors by their parasites. *Annual Review of Entomology* 48: 141–161.

Hurd, H. 2005. Parasitic manipulation: Stretching the concepts. *Behavioural Processes* 68: 235–236.

Kavaliers, M., and Colwell, D. D. Aversive responses of female mice to the odors of parasitized males: Neuromodulatory mechanisms and implications for mate choice. *Ethology* 95: 202–212.

Kavaliers, M., Colwell, D. D., and Choleris, E. 1998. Analgesic responses of male mice exposed to the odors of parasitized females: Effects of male sexual experience and infection status. *Behavioral Neuroscience* 112: 1001–1011.

Kavaliers, M., Colwell, D. D., and Choleris, E. 2000. Parasites and behaviour: An ethnopharmacological perspective. *Parasitology Today* 16: 464–468.

Klein, S. L., Bird, B. H., and Glass, G. E. 2001. Sex differences in immune responses and viral shedding following Seoul virus infection in Norway rats. *American Journal of Tropical Medicine and Hygiene* 65: 57–63.

Klein, S. L., Bird, B. H., Nelson, R. J., and Glass, G. E. 2002. Environmental and physiological factors associated with Seoul virus infection among urban populations of Norway rats. *Journal of Mammalogy* 83: 478–488.

Klein, S. L. 2003. Parasite manipulation of the proximate mechanisms that mediate social behavior in vertebrates. *Physiology and Behavior* 79: 441–449.

Klein, S. L. 2004. Seoul virus infection increases aggressive behaviour in male Norway rats. *Animal Behaviour* 67: 421–429.

Koella, J. C. 2005. Malaria as manipulator. *Behavioural Processes* 68: 271–273.

Kuris, A. M. 2005. Trophic transmission of parasites and host behaviour modification. *Behavioural Processes* 68: 215–217.

Lafferty, K. D. 2005. Look what the cat dragged in: Do parasites contribute to human cultural diversity? *Behavioural Processes* 68: 279–282.

Maitland, D. P. 1994. A parasitic fungus infecting yellow dungflies manipulates host perching behaviour. *Proceedings of the Royal Society of London B* 258: 187–193.

Moore, J. 2002. *Parasites and the Behavior of Animals.* Oxford University Press, Oxford.

Moore, J., Adamo, S., and Thomas, F. Manipulation: Expansion of the paradigm. *Behavioural Processes* 68: 283–287.

Moshkin, M., Gerlinskaya, L., Morozova, O., Bakhvalova, V., and Evisikov, V. 2002. Behaviour, chemosignals and endocrine functions in male mice infected with tick-borne encephalitis virus. *Psychoneuroendocrinology* 27: 603–608.

Mouritsen, K. N., and Poulin, R. 2003. Parasite-induced trophic facilitation exploited by a non-host predator: A manipulator's nightmare. *International Journal for Parasitology* 33: 1043–1050.

Nickol, B. B. 2005. Parasitic manipulation: Should we go anywhere? *Behavioural Processes* 68: 201–203.

O'Donnell, S. 1997. How parasites can promote the expression of social behaviour in their hosts. *Proceedings of the Royal Society of London B:* 264: 689–694.

Poulin, R. 1992. Altered behaviour in parasitized bumblebees: Parasite manipulation or adaptive suicide? *Animal Behaviour* 44: 174–176.

Poulin, R. 1995. "Adaptive" changes in the behaviour of parasitized animals: A critical review. *International Journal for Parasitology* 25: 1371–1383.

Poulin, R. 2000. Manipulation of host behaviour by parasites: A weakening paradigm? *Proceedings of the Royal Society of London B:* 267: 787–792.

Poulin, R., Brodeur, J., and Moore, J. 1994. Parasite manipulation of host behaviour: Should hosts always lose? *Oikos* 70: 479–484.

Poulin, R., Fredensborg, B. L., Hansen, E., and Leung, T. L. F. 2005. The true cost of host manipulation by parasites. *Behavioural Processes* 68: 241–244.

Rigaud, T., and Haine, E. R. 2005. Conflict between co-occurring parasites as a confounding factor in manipulation studies? *Behavioural Processes* 68: 259–262.

Rothschild, M. 1962. Changes in behavior in the intermediate hosts of trematodes. *Nature* 193: 1312–1313.

Sapolsky, R. 2003. Bugs in the brain. *Scientific American,* March.

Sheffield, C. 2000. Nobody tells me what to do. www.fenrir.com/free_stuff/columns/science/sci-044.htm.

Thomas, F., Poulin, R., Guégan, J-F., Michalakis, Y., and Renaud, F. 2000. Are there pros as well as cons to being parasitized? *Parasitology Today* 16: 533–536.

Thomas, F., Schmidt-Rhaesa, A., Martin, G., Manu, C., Durand, P., and Renaud, F. 2002. Do hairworms (Nematomorpha) manipulate the water seeking behaviour of their terrestrial hosts? *Journal of Evolutionary Biology* 15: 356–361.

Thomas, F., Adamo, S., and Moore, J. 2005. Parasitic manipulation: Where are we and where should we go? *Behavioural Processes* 68: 185–199.

Tompkins, D. M., Mouritsen, K. N., and Poulin, R. 2004. Parasite-induced surfacing in the cockle *Austrovenus stuchbury*: Adaptation or not? *Journal of Evolutionary Biology* 17: 247–256.

Webster, J. P. 2001. Rats, cats, people and parasites: The impact of latent toxoplasmosis on behaviour. *Microbes and Infection* 3: 1037–1045.

Webster, J. P. 2005. Parasitic manipulation: Where else should we go? *Behavioural Processes* 68: 275–277.

Wellnitz, T. 2005. Parasite-host conflicts: Winners and losers or negotiated settlements? *Behavioural Processes* 68: 245–246.

Wilson, K. 2000. Parasite manipulation: How extensive is the extended phenotype? *Trends in Ecology and Evolution* 15: 395.

Zimmer, C. 2000. Parasites make scaredy-rats foolhardy. *Science* 289: 525–527.

INDEX